The Purple Book

Mark Mozart

Copyright © 2012GriffNotes Publishing, A Griffin Cleft Subsidiary

All rights reserved.

ISBN:10-1540731162
ISBN-13:978-1540731166

DEDICATION

Griffin Cleft, dedicates this book to the puzziling memory, and legacy that is
Marcus Griffin.

A life only lived, until age Nine, Marcus quietly struggled with Autism. A struggle unnoticed by Marcus, as he left us with an abundance of brilliance, despite his inability to speak. A greatly impressive young man. We heard you loud and clear.

A donation will be made to the Marcus and Friends Charity, benefiting the families of, children with Special Needs. Thank you for supporting the effort to bring more awareness, and resolve to the issues that regard Special Needs Learning Disabilaties.

We all love and miss you, Marcus....

CONTENTS

	Acknowledgments	I
1	The Purple Period	1
1A	Requiem for the Struggle	3
2	God Bless the Child That Got Its Own	14
3	Mark, Set, Go	25
4	Standing With Truth	41
5	Good Help Is Hard To Find	56
6	Takes One To Know One	80
7	Seven	90
8	Jesus Take The Wheel	101
9	Black Lives Matter	117
10	Worlds Apart	151
11	Coming and Going	185
12	You Win Some	219

ACKNOWLEDGMENTS

I began writing the Transistion Memoires with a plan to share the stories of very specific transistions beyond death. However, Omhmar Griffin, did not live as long as we anticipated. And, unfortunately could not remain to write the conclusion to his transition. As a result, the book that you are about to read has not been edited in anyway, to honor the realtime trnsistions, written about in our memoirs.

We ask that you forgive us, and offer us an abundance of patience as you stumble upon disruptive, typos, and or mistakes.

1 THE PURPLE PERIOD

At least, I believe! I believe that it will be a physical event! A physical manisfestation, and not another metaphoric offering of fables. Not fictional, Gospel Fables that allow me rudementary comprehension, since. I am a dumb ass Human. I should be running at frightening speeds, with mine eyes lifted toward the hills. I should run away from their sounds of opression. I hear their scheme to keep me opressed. Call me crazy if you wish, but at least I believe. I believe in Eternal Life!

God is so gentle and seemless in manuvering my transitional periods. Once I discover that I have transitioned to a new point. He's always there to whisper in my ear, "It should only feel like the next logical step." If there is to be some major difference between the two words. I would have to say that God's Process is a process of logic, while his son's process is inately practical. Yet,I think that he has overlooked praticality in his askings of me.

I'm learning that the system is trully rigged to promote failling, instability, struggle for the participants that fall to the necessity of entitlement programs. When I would hear older cats hollering about the "Man" keeping us down. That our own Governement was trying, with full force, to maintane a status quo inclusive of check points that enssured that we as a broke community, would surley be left behind. I thought those to be convenient excuses to why they always fail. But, I don't any more.

My elegant, career afforded me wonderful opportunities not normally appreciated by a large majority of the world. Especially not, those trapped in their inner city oppressions. As delusional as I lived in my bubbled world, lacking of all the evidence to support the claims of the men that seemed to me, to be just cry babies. Punk as bitches that felt they were supperior to me because, I was the faggot. Meanwhile, I have been working, top industry gigs since I was Sixteen years old. Hob nobbing with their favorite celebraties, gracing the stages of Theaters and venues that they will die without stepping One foot in.

I was debt free, and never had to accept a dime from the instituition that has funded their entire lively hoods, as adult men.

Granted I have spoken my choices, and aggreed to commitments of what works best in my life. But, I believe God and his son have designed my journey, along side me. God, Jesus, and I arrive at this point in our story. A story often inclusive of battles that gain and loose me. This time, this battle. I get my chance to experience how niggers are truly treated in the Black Ghettoes of America. Just for a short while. And, the battle is not really mine. I call this our Purple Period.

1-A REQUIEM FOR A STRUGGLE

Just because I fell victim to harsh economic down turn. I would now be forced to live in places where crack heads and pan handlers would not only know where I live. But, would also knock on my door to invite me to smoke crack? Yes I did in fact try it twice. The curiosity in me got the better of me. I don't know why I can be obsessive about disputing the theories I hear. And, a drug that attracts an abusive addiction after only doing it One time sounded very odd to me, in fact.

Not only did I not enjoy a single feeling from the high, I never thought about doing it again after that second time, that I touched the stuff. But, yes. Chillin with the same bums that would stand at the stage door as I exited to receive the single dollar that I would toss their way each night. Who would have thunk it? But, wait it gets better.

I have been stopped and frisked, and demanded to produce an identification at the officer's beckening, Three times. Which is highly illegal, for anyone else who is not a young, healthy Black man. I twice, have been gay bashed by a neighborhood, hatefull ass thug, that has huge issue with how I choose to live my life. Thus creating entire blocks within my hood that I can no longer frequent.

And, now aressted and injured for tresspassing in my own apartment. Don't even get me started on the lack luster dating that goes on in the hood. With all of us too stressed for making ends meet, to make true attempts at romance. We're all only half present, while we can only afford to fuck!

Get your head out of the clouds! Heard that one before? Cliches always seem so old and familiar. In whatever language. The fatal results that can manifest when getting lost in the cloud only conceptionalized in 1927. A hazard single handidly introduced by an American Hero, Mr. Henry Ford.

The Model-T gave America a great convenience, yet required new levels of multi-tasking, and concentration American's had not yet practiced. Not, just when driving them on the road. But, also when building them on the assembly line. Thus, get your head out of the coulds, became the new "Watch out!" Warned to us daydreamers when not paying attention.

I was an Honor Roll student until I got to high school. Placed in AP level Regents courses in Junior High School. When choosing a high school, I was certain that I would get into Stuyvesant HighSchool. Not realizing that the entrance exams would include mostly information on them, not offered to public schools in districts populated mostly by imigrants, and minorities. So, when I failed miserably at all of the Science Exams, and application deadlines for any other New York City, Specialized High School. I was forced to spend my Freshman year at McKee Vocational and Technical High School.

Which was terrifying in itself. Since, most of the kids in my freshman class, were the bullies and ratchet kids from my junior high school. It wound up not being that bad. Once my classmmates saw what a star I was on the Track team. I earned the same respect that a Varsity jock would be due. My exploration of life really took off at McKee, thanks to Ms. Smith.

She taught English, which was my favorite class of every day. Only because she made it her priority to design and execute a curriculum that would be identifiable to Black kids our age. She'd crack me up at how her definitions for our weekly vocabulary words could never be found in any dictionary on earth. The word that spoke with me the most was Cliché. She, claimed that it was the word describing an overworked expression. I often wondered how does an expression become overworked. I'm clear now, thanks to Reality Television, and it's stars such as Ne Ne Leaks', "I'm Rich, Bitch!"

There is also an everlasting need for the use of cliches, after they've entered the Status Quo. Isn't there? Blame it on history repeating twice a century, or the oppression that antagonizes human evolution at such a slow progression. Just think. It would be Two Thousand and Nine years before the Free World would feel comfortable enough to let someone other than an Old White Man run it. Ending a civil rights movement that would span decades. Even centuries, depending on when you start your timeline. However leading American Civil Right's Leaders who were hangging on to dear life. Just long enough to bare witness, to say, "Well, I can die now. I've seen everything."

Most everyone that knows anything about me has always called me a "Big Dreamer." When you spend your childhood watching the promos for "Cat's", the Broadway Musical. Wishing to be a part of the cast. And, then not only grow up and perform in "Cat's" but, also land the role that is featured in the "Cat's" promo I grew up watching, "Mr. Mistoffelees." I'd have to say that, quantifying the magnitude of my dreams, are mysterious equations.

Twelve years into the Mellenium we've finally entered the teens. I spend the first decade on the West Coast of America. Now, 2013 is here and I am back home in New York City. The Thirty year old having to move back in with mom. Whether it's because you have spent the last decade fucking up in life or like in my case. Mom has fucked up her own life so much that in her young senior years she needs someone to take care of her. At the point where my turn to take care of may ailing "Baby Boomer" mother arrived, I had just turned Thirty and had been a professional dancer for Fifteen years.

I trnsfered my sophomore year to the High School of Performing Arts, and began my training as classically Ballet dancer. When all of my high born friends from junior high went on to Stuyvesant without me. I couldn't see celebrating a career inclusive of film, stage, and Television in my future. I arrived late into my training. But, I did not wait long to start working. You can guess what my first audition was for. Yup, Cats! But this isn't a story of how easy my life or anyone else's can be. I wish I could tell you that I was the wonder kid who won his dream job, at a young age, and after prematurely attending his very first audition.

No, that would be Jeremiah. Jeremiah had not arrived late into his training.

He was fortunate enough to be blessed with a mother that quit her jobs at times. In effort to travel to a foreign city, so that her son could pursue the quality trainning that only a school like San Francisco Ballet would offer. The trade off is that, by age Sixteen. Jereiah would become the youngest dancer ever to be accpeted into the Alvin Ailey American Dance Theater. It would be years later before I would be landing the role of Mr. Mistoffelees.

Years of Ballets, Operas, small roles in T.V. and Film, trade shows, cruise line shows, fashion shows, background dancing for Recording Artists. Many years of grinding the hustle that pieces enough gigs together in a year for a dancer to feel like a full time professional. Of course seeing the world on your employer's dime is a fabulous perk to the industry. But, no gig that I landed seemed to be big enough for me. I wasn't simply dreaming. I was after something here. I'm talking major warpath. I wanted to be the Black Dick Clark of our time. Unfortunately, I think Nick Cannon beat me to it.

Like star atheletes, Ballet Dancers retire early. Most of us usually have a second career either in education, or in entertainment production. After visiting home for Christmas, and witnessing the horrific health condition my mother was in. I knew it was time to move on back. So, in 2013 I retired from dance. I hung up my pointe shoes. That's an overworked expression we say in the ballet world when we retire. It was a difficult decision to make, but I did not know how long it would take to return home to Los Angeles.

Funny enough, I never even wanted to move to Los Angeles. I only went because every Casting Director in NY told me that I looked to young when I was auditioning as a kid. And, advised me to move to Hollywood where film and Television casting needs are younger. We all agreed that after I aged a bit, I could move back to NY and continue the dream of becoming a Broadway star. Now that I am back in New York. Becoming a Broadway star is no longer the "Big" dream. After working as a dancer for years, the idea of performing the same show eight times every week, week after week just seems like grueling work. No. The new "Big" dream is to use the many talents of my family, to create lucrative family empire. And, that dream would have to wait. At least long enough for me to help with stabilyzing Mom's health.

I didn't grow up in the fun fast-paced New York City that you see on T.V. Staten Island is New York City's Miami. Where only old people and White suburban families mostly reside. Not much eye candy to salovate over. The Italians that rule the Island are drop dead gorgeous when they're too young to do anything about. Yet, they don't age well at all. I think it's all the pasta sauce. Not many single people either. We all move away after high school and return once we're married. Not married yet, so back to mom.

If you were the type to leave home right after High School, moving far away. Only to come home for a Christmas visit once a year. Then you might relate to missiong out on seeing your family's evollution without you. Your nieces and nephews who were once babies in yor arms are now teenagers, or adults. You knock on the door one Christmas and your deceased Grandmother opens it. "Scream!", I let out. "Boy what's wrong with you?" our mother says. "Just noticing that you look how I feel", beat down by the devil. One foot dangling out the coffin. Government assisted. I'll move on, since you get my drift.

The past year I'd managed to save up some money, and I still had gigs lined up. So, I wasn't financially strapped. I was, however, emotionally drained since I had just disbanded the Ballet Company I began in 2003 with my best friend. The Los Angeles Modern and Ballet Company was a dance company that Jeremiah K. Tatum and I began. Even though we were both young and still dancers ourselves. We wanted to create opportunities for women of color to perform Classical Ballet. Closing the company wasn't an easy decision. The strain of juggling business and pleasure was taxing on the relationship that Jeremiah and I shared.

I met Jeremiah when he was Fifteen and I was Eighteen years old. Immediately I fell in love. He was underaged and I couldn't let on that I was smitten. Or, that I had never in my life met anyone as incredible as he is. Anyone as open minded, or talented, or as impressive as a teen. When I say impressive, yes. The boy danced better than anyone alive to date. But that's not what I am talking about. I mean how The Lord works in mysterious ways. Why would he package the epitome of perfection into the most spectacularly sculpted body. With the prettiest chocolate skin. The highest, roundest bubble booty, that doesn't distract from the symetry of his horse like legs. And then put that body on a kid? Worse, put that kid, with that body, in tights? And, then bend him over at the Ballet Barre right in front of you?

Naturally my plan at first was to pretend to hate him. Remember I was only Eighteen. We were dancers in the Lula Washington Dance Theater. Dance companies can be kind of clickish. Kind of like a real housewives show. And, sometimes for very arbutrary reasons like Talls against the Smalls. Lula's clicks were divided in the usual way. The company members who trained at the school against the company members who trained elsewhere. Unusual to me, was a subset of Holy Rollers. Common to most Black Dance Companies in America, is the click of Christians. The click that prays and worships together before a performance. While judging the rest of us, as we go to the club after the performance, leaving them alone in the hotel. This came as a surprised to me. Although trained in all forms of dance. I had only been a Ballet dancer until now.

The cultures between Modern Dance and Classical Ballet companies are quite different. Jeremiah was a student of Lula Washington's while I was a graduate of Laguardia. Which is the school that pops out a handfull of Hollywood Divas every year. So I had no problem throwing him divalicious shade. Our fights were epic! I think that everyone in the company could feel that they were mostly fueled by sexual tension. Tension that relieving would nail my ass under the jail.

But, there was one company member, a girl part of Jeremiah's click, who just couldn't accept the fraudulence of our conspiracy. She fell for the facade of our arguments every time. She just couldn't be convinced that Jeremiah was well on his way to the Castro. If I even looked at him too long, she was eager to give me the stink face. She even organised a grand event where she and Jeremiah would double bank me. Not the good kind either.

After just landing in Pheonix, I could tell that something was up. Jeremiah was being extra nice to me, while the rest of the company were whispering nothings all day. We checked into the hotel just long enough to brush our teeth before heading to the theater for a technical rehearsal. Since Lula Washington Dance Theater has the longest tech rehearsals of any dance company on the Planet. Tech rehearsals are supposed to be easy breezy for dancers, and more focused on the technicians. But, not for The Washingtons. She not only wanted you in full costume, but also dancing full out as if a packed house of ticket buying patrons were present.

After rehearsal, the nice giuy act came to a screeching halt. I contributed Jeremiah and old stink face's bad energy to the fatigue of starring in twelve ballets on a single bill every night. But, the whispers from the other dancers have been floating all day. I didn't get suspicious until the road manager paid a visit to my room after the rehearsal. "How are things between you and Jeremiah?" She says to me. I'd already been working professionally for Six years, so I know how company politics worked. By the time the pety squables between company members reach management, somebody is about to be fired. I'm sure it ain't the one who just arrived. "Ok, I'll talk to him."

And, that was just what I intended to do. But, as I am calming self down, and awaiting his answer of the phone I hear the ring interupted with her voice. Yup, old stink face. She didn't miss a beat. As if she knew that I would be calling at that exact moment. Remember that hotels didn't have caller I. D. back then. At least, not the hotels that second rate, broke down Black dance companies stayed in. But, some how old stink face knew that it would be me on the other line. Calling to make peace with Jeremiah, and that she wouldn't get to host her WWF Smackdown event that she had been planning for weeks. She wasn't about to let no bitch ass, upper management mediation ruin what could be a pay-per-view worthy event. She had already chosen the venue, her outfit, bell time, and sent her invitations out. With all the R.S.V.P's she received? It was going down.

You'd think that having Three years on the two of them would have positioned me better qualified to difuse the situation. Nope. I told them that they could meet me in the hall right then, since our rooms were across from one another and slammed the phone down. As I flung open my door, I see that most of the company had already taken their front row seats. Who knew? I had just agreed to old stink face's venue and bell time. Ready to whip both their asses, Jeremiah unfiles from behind her and it becomes like the dance at the gym scene from West Side Story.

A spotlight unfolds, illuminating his body I described earlier. Everyone else in attendance suddenly disapears. As our eyes lock I wanted to lift him up and do a slow spin around in a circle as if I were Tony and he Maria. I could tell that he wanted me to as well. And, as the sound of our cheering crowd swells to it's full volume. I have visions of hand cuffs, and jail cells that grow larger and slowly fades toward me. Finally someone grown enough to respect the liability of Lula Washington's insurance intervened and old stink face had her main event cancelled.

Thank God, too. I don't know if Jeremiah and I would have ever grown to where we are, if our relationship began with domestic abuse. The next city after Phoenix we dropped the fake beef and started getting real. It was very hard for me to become his friend because our chemistry and sexual tension only intensified.

Once we returned to Los Angeles, our childish dates mostly consisted of sitting in my car listening to music and talking whenever we could. We'd sneak to the shadows of some dark street in his neighborhood as I would drop him off after a performance or rehearsal so his mom wouldn't see us. We kept it purely innocent. But, the image of a grown New Yorker spending hours alone in a car doing anything with her son would cause problems for said son.

Jeremiah's taste in most everything was ecletic to say the least. As eclectic as he was, he still managed to have really relevant posessions. I felt so uncool around him. He wore really expensive designer clothes, and could name every model and designer photographed in Fashion magazines. Which, I never read. And, he knew all the words to Indie albums that Black guys didn't ever listen to. Who's ever heard of Tori Amos before? I didn't.

Just as I was falling in Love with Teenage Suicide, and secretly Jeremiah too. He got the call from Ms. Judith Jamison to join the Alvin Ailey American Dance Theater. Not a minute too soon. I was spared hand cuffs and jail cells and he became the youngest dancer to join Alvin Ailey. More about him later, Mom needs me.

"Mom what's wrong?" I ask when I notice her eyes filling up with worry. She seemd fine a few minutes ago as we took our evening stroll aroud the neighborhood. But, now here she is grasping for air and not able to get much sound out to respond to my inquiries. I quickly grabed the phone and dialed 911. When the paramedics arrived I felt as though I were the worse son on the planet.

We're talking about Staten Island here. As usual, two arrogant and judgemental White men wishing that their jobs didn't require them to ever touch a Black person. Takes their time entering our home as if it were not an emergency. And were more concerned with why I did not know which medications my Mom was on than saving her life. I'm new at this being a son who lives with his elderly parent thing. So, I didn't know that I should know information like last doctor visits or prescribed medications. But, I did know that she was allergic to shell fish. As I told them this, one of the paramedics dropped a suringe he was preparing to inject something into my mom with. As I watched the needle pierce into her carpet, I realized that an allergic reaction to anything when she already can't breath and suffers from panic attacks, would have surely killed her. A bitter sweet truth. But, I was happy that it was my turn to save her life.

Since the doctors have decided to keep my Mother for observation. I decide to take a short break and go down south. Both of my sisters live just a short drive from Staten Island, near Baltimore, Maryland.

Spending so much time away from home, I didn't get to see my teenage sister become a woman. She now has two boys of her own and holding down quite the suburban life with a gorgeous Three-bedroom Townhouse and a dog.

My oldest sister and I have grown closer in recent years. Last year I helped her bury the last of three sons who never even made it to becoming even teenagers. I don't know why I didn't get it with the first two. However, The third time is the charm. With the death of my nephew, Marcus. I finally realized the blessing that each day given, actually was. Not a single day is guranteed by your birth on Earth. Everyday is another day that you can add to your story. Your legacy. Your purpose. I know that you hear people saying that all the time. But, it really hits home when you witness a loved one fall to pieces as undertakers lower her Nine year old child into the ground.

He'd been diagnosed with Autism since he was Two years old. At that time we knew that the road for raising him would be an unpaved freeway. However, we were optimistic that he'd at least make it to college. As you can imagine, him not even making it to Junior High School was a shock to all of us. You want to meet a strong Black woman? The experience of burrying a child for the third time is enough for me to question the validity of worshiping Christ. But, she was in Church the very next week. In fact, I assumed that I would mostly spend my visit to Maryland consoling her. She and I wound up ministering to one another the entire weekend.

I had already left the Christian Church in 2005, after being a faithful follower all my life. So, I really wasn't trying to hear no Jesus preaching. But, I think you'd agree that her tragic lost yielded the obligation to at least pretend to listen. In an interesting turn of events, moments of ministry took a revilating journey through our very similar belief systems. Her's based in Christianity. And, mine based on the teachings of Abraham-Hicks.

Whenever I introduce Abraham explanations of The Law of Attraction. I invite them to select any video of Esther Hicks, chanelling Abraham on the internet. I do this because I find it to be the best way of introducing the basic concept of the Law of Attraction. What ever big question a person is toying over when I have them chose the video, the video usually offers them clarity.

This was not the case with my nephew's mother. In fact it sent her running up the stairs. "Hold on Omhmar. Wait right there." She shouts as she books up the stairs. Omhmar was the name given to me by my father when I was born. But, bailed before I was old enough to learn how to pronounce it. So everyone just called me Omar. Which my Mom adamantly disputes is not how my Father pronounces it.

"Where are you going?" I ye'll back. "I gotta show you something." I hear her say from the top of the stairs. As I wait for her return back to the kitchen where we were having "Church." I think to myself, how much my oldest sister had grown. There use to be a time where she dominated conversations so heavily, I wouldn't be able to get a word in edge wise.

But, here we are introducing new ways of thinking to one another. I'd already been familiar with the lessons of Jesus Christ, but never put two and two, together to equate that what Abraham is teaching echoes alot of What Jesus tried to teach. "Let me show you this!" She's back and she's excited!

I promise you that I did not manipulate anything on my computer to make her choose the video that she chose. She chose a video from a seminar where one of the attendees asked Abraham a question regarding children with special needs like Autism. Abraham's response described a paradigm shift in our world. A shift where new Humans are requiring us all to communicate in ways contrary to how we were use to. After realizing that the video was the exact validation of ideas she'd been recently questioning. She ran up stairs to get a legal sized manilla envelope. On the outside she had written in Black, bold letters, "A New Paradigm."

Before Marcus died, my sister had been compiling literature that presented ideas relating to communicating with non verbal, Special Needs Children. New techniques, such as Picture exchange. An exercise, where the children select photographs, such as their favorite cereals, to a family member they might want to visit. The pictures hang stationary where the child has access to them at all times. When Marcus would miss my Nephews, his Cousins. He would grab the picture of them off the board and bring that picture to my sister. This process alowed more specivity in our family knowing what Marcus' immediate needs would be.

Techniques such as this and many others had inspired her to enroll in school. Having Marcus, prevented my Sister from finishing up her degree, just a few credits shy. But, now at the end of her journey, raising Marcus. She is convinced that finishing her degree and working with Special Needs children is her logical next step.

2 GOD BLESS THE CHILD THAT......

Happy to see Mom's health stabilizing. That health scare that landed her a week long vacation in the Student Hospital has put fire in her belly. I'm starting to see many improvements in her overall mobility. Instead of falling asleep holding her Bible and cross, as if she awaits the Devil in her sleep. And, instead of falling asleep to TBN and Praise The Lord, and pastor Crouch or Kenneth Copeland. She now falls asleep to Scandal, or the Wendy Williams Show.

Now before the "Church" accuses me of causing mom to backslide. You really must understand the fear that consumes Mom on a constant bases. All of my life I have watched her read her Bible cover to cover, multiple times. As good of a read as it may be, anyone who's attempted reading it in it's entirety surely understands the time committment it stresses. Our relationship is also improving. It's great to get to know her as an adult. And, introducing her to the new me. Name change and all.

One good thing about falling below the poverty line is that I now qualify for entitlement programs that I never did before. And, perfect timing because for some reason, my unemployement checks from LA have stopped coming. And movie residual checks get smaller and smaller. Although, I am still very happy to see a SAG check sitting in the mailbox. I'd enrolled in a housing supplement program that doesn't seem to be very popular amongst landlords in New York.

I never thought that I would live in New York again. Especially, not this new New York post Guiliani. But, with my siblings all down south. And, with their own families to tend to. I'm the only one left to take care of mom. It was actually my plan to just give mom a year, and then I would move to Indidanapolis, to start my family and find a husband. But, since our reunion has turned into a sweet one. I've decided to give mom another two, years. And, hope not to ever have to go through the grueling, Amazing Race that New Yorkers call an apartment hunt!

Just before mom went into the hospital, and I visited my sisters in Maryland. I submitted all the final paperwork for an apartment I couldn't get denied for. I was in adherance to every compliance required for approval. That's in addition to finding a landlord willing to wait the three-week process before getting an approval. I was sure that I would return to the good news of a brand new apartment, ready for me to call my own. But instead, I was denied once again.

The first time, I was denied because it was not a two-year lease. Now this time, I'm denied because although I originally aproved for a one-bedroom, new laws have been implemented during my search period. I now qualify for only a studio apartment. I know that beggars can't be choosy. But, I am concerned with the size of the apartment allowing my plans to bare creation. But, more importantly I learned that the City of New York does not consider anyone who hs a regular couch to sleep on every night, homeless. So it's off to a shelter I go.

I'm horrifically terrified of rats! Hilarous when you think of how many times I've danced the role of the Rat King in Tchaikovsky's The Nutcracker. But, my mom, younger sister and I spent some time in a shelter before moving to Staten Island. It was brief, but fixates terror in my memory from rats that are ride or die, really sticks with you. And, certainly painted the picture in my mind that New York City Boroughs that were not Staten Island would be rat infested.

So as I moved into West 137th street, I was concerned that I would have furosious panic attacks from the rat infestation. Happy to say, that I didn't see a single rat my entire time there.

As I move in, I realize that it has been a full year and two denials since I began looking for an apartment. Praying that this move is the last step to finally being approved, we may begin our journey.

Valentines Day 2014. I am really starting to get the hang of this Law of Attraction. Yet, I still can't seem to attract a decent lover. I am sure that I don't want to date anyone seriously right now. But, I would like to have a fuck buddy to kick it with from time to time. Instead of having to spend most of my horny hours, searching and interviewing on line. I can't actually have company in my room. Not to mention that I have been on mom's couch for more than a year, and for the first time since leaving LA. I can be alone in my own.

Abraham-Hick's Law of Atraction is not a religion. But, after spending time with my older sister, we both realized that accepting the Law (s) of Attraction can make you a better Christian. Or, whatever you are. No, seriously.

For example. If you beleive that you have a father in Heaven who watches over you and your best interests. It is so. Now whether or not he fathered Jesus Christ as well becomes erelavant. Following the blue print so eloquently laid out for you in the bible, not only creates the since of a Heavenly bond between you and your father in Heaven. But also, provides the evidence that gives confidence to the idea that you have developed something unique in special. A relationship, that since rooted in Chritianity, would mirror that of the relationship between the original Christ and his Father.

But, I'm not Cristian yet. well, I was..... At any rate…

I agreed that I would return to church. Not to take Christ as my Lord and Savior. But, to fulfill my responsibility to participate in the study of where we came from and where we are going. When I left the Church, I was very angry. Confused, but mostly angry. I could see within my best friends, Jesus working in their lives. But, I could not see him working in my own. I mean, one friend! She literally would say, Father God, I need some new blinds for my apartment. And, the very next day a neighbor would knock on the door, asking her if she wanted the blinds that the neighbor was throwing away.

My mom would call that God's annointing. She's always claimed that I was annointed as well. But, I've never felt that I was. Some people call it ungrateful. While others call it big dreaming. And, I call it bamboozled.

How is it, that I go to church every Sunday? Pay my tithes and offering? Live Christ-like during the week? Fellowship, volunteer, and extra prayer and Bible Study throughout the week, and. Be a good neighbor and shuch. I still never seem to see my gifts come back Ten fold. For about the last Three years of my last attempt at being a worshiping Christian, I only prayed to Thank God for what I could see as blessings. What's with the whole if it's not in his wil for you, the answer's going to be no? So, I didn't find any pupose in praying to ask for anything.

Since leaving the faith, I've not left behind my spirituality. I am really beginning to master the practice, as well as enjoy the results from the Law of Attraction. If you are unfamiliar. Esther Hicks channels a collective, that calls themselves infinite wisdom. They channel Esther from a placed called Source Energy that she has named Abraham. I'm not trying to pursuade you into any spiritual, political, or religious direction. This book tells the story of how Omhmar and Jeremiah equals Mark. The three of us, are working closely with God in an effort to make the world more peacefull. The Pruple Book tells the beginning of our journey away from the America that hates me for being Black and Gay.

So here we are at West 137th. As Cupid's arrow gathers lovers on the street beneath my window, I could hear the sounds of lover's quirells and salutations. I couldn't have been happier to be alone. Although I only had my blackberry and note pads for insistant planning and product development. I was eager to pick up where I left off. Now....I wasn't sure where I could pick up from. Of course, I'd love to say, let me get back to the status I enjoyed when first arriving to New York, with success to boast, pockets full of savings, and resources galore. Unfortunately, a little crawling was needed.

Up until now, I'd had a lot of experience analyzing dreams. Also, when people who are close to me have a mother pass away. She, usually tries to communicate with them through my dreams. It's worrysome activity, but I think it has something to do with me being a "Momma's Boy." So right away when I get to West 137th, I am aware that the energy is always too full around me for me to be alone. I began to hear voices. Or at least, I thought I heard voices.

"Dr. Kirkland, I think I need to see a Mental Therapist." I blurt out, in my confused pondering face. "I can write you a referral, Mark. But you wanna tell me what's going on?" Dr. Kirkland asks. Sexy, Jamaican Dr. Kirkland. "I am hearing voices. I'm not scared of them. They actually offer excellent guidance." "Do you see the people that are talking to you?" He asks. "No" I reply. "Well I don't think that it's psycosis." He says. Explaining to me that if I were to start seeing the source of the voices, we'd have more issues. After, seeing the head of the Mental Health Clinic, I was assured that nothing abnormal was happening. Or at least they attempted to assure me.

Yet, I knew at the very least Abraham was trying to contact me. I couldn't understand why exactly. The messages that I was hearing were all positive affirmations with this underlying mantra often being chanted, "You can't get it wrong, you can't get it right." And, with that in mind, I set out to take over the world. Starting with a reasonably priced One bedroom apartment located on an express stop, and in a decent neighborhood.

As the work developed. I began creating viable scripts to become entertainment products, such as movies and concerts. With nearly finished products, I began to really apply the Law of Attraction to my life. On a side note, laugh out loud. Jeremiah is in town doing his usual Broadway gig. Wouldn't you know that the first night he is town, our paths would cross in Chelsea. And, then again the next day in Times Square. And, once again the next day in Central Park. How can two guys, who supposidly, once upon a time loved each other deeply, now walk right past each other without saying a word?

I would love to continue this childish, foreplay. But, lucky for him his Birthday has arrived before mine, and my mom raised me right. So, of course I fold. "Diva!" With a big fake smile, I say. He responds, in the only way we know how in Los Angeles. "Hey Bueen!" First we call you by our signature pet name for the familiar faces whom names can't be recalled. Secondly, we pretend as if we didn't just circle back, passing eye contact four times before finally acknowledging one another. And, lastly you act as if there is no ill feelings harbored from past events.

My financial stability was improving to the point where I could afford to show an old friend visiting my fair city a good time for his Birthday. Later that night we did have a conversation to try and ease things. However, he'd just enetered the age where he has to prove his independance. Be it, living, freedom, or perception. And, I'd entered that age where I just didn't give two fucks enough to debate to the death of him. So, as he returned to Los Angeles. We returned to what was familiar behavior on his part. Out of sight, out of mind. I couldn't get him to answer a call nor reply to a text.

The break up of our Bromance, is really just an extreme case of miscommunication. The last we spoke, I had told him that his mom was getting on my last nerve. For years I had allowed Jeremiah to use me as blame for whenever he needed to get out of trouble with his Mom. So naturally she'd developed the perception of me being a bad influence on her so. Not nearly the case! Jeremiah is actually the bad influence on me. While accepting his blame, it never occurred to her, that Jeremiah was actrually the one always in trouble. Requiring alibys and shit. Meanwhile, I was the one holding the full time job, a long term relationship, my own gorgeous apartment, and building several companies.

However, the hand full of negative stories I actually have contributed to his famiy's common knowledge. Are the stories she chooses to tell, publicly. I can handle a little embarrassment. What I can not come back from, is when his Mom attends important galas, sits across the table from a well respected Arts Presenter, whom has never heard of me. And, tells him quite a bit of my own personal business that should never leave the kitchen table talk. So now, when I apply to be presented by said presenter. Jeremiah's Mom has already given the presenter my first impression. And, not a good one.

As the news of her public bashing contnues to come my way. I simply asked him if he would ask his Mom to stop hurting the efforts that he and I were making in the classical arts. I quickly understood where he was coming from when he said, "That is my Mom. It really pisses me off when you come at me like this regarding her. What do you want me to say to her. You need to shut the pepole down who are messaging you the hurtful information, so that you don't hear about her bashing. That's where your problems start. Just like I shut you down whenever you come at me sideways about my Mom, do the same to them. We don't want to hear it." He means he and I when he says we. But still.

The nerve of that little fuck face! I hate to be silenced in a way that reads me absolutely correctly. At this point we have both grown up from the little Mommas Boys that we were once were. I had my own relationship with his mother and could have just as easily gone to her myself. And, expressed to her the damage being done, by bashing me in certain circles. But, that doesn't erase why I am mad at him. After nearly Four years of not speaking. I couldn't believe that he thought it was over the bullshit with his Mom. I even apologized for my comments and told him that it was squashed.

My grievance occurred just weeks before this incident with his mother. Jeremia asked me to join him in Memphis for a short gig that he'd booked. It was not uncommon for me to visit a city that he was performing in, just for the purpose of being his travel companion. When he begged me, I said yes. I didn't want my pumpkin to be South of the Mississippi, all alone without propper supervision and witnessing of his night time affairs. We both can turn up pretty well. And, if they are to shoot a young brother. At least I might live to tell hismother the truth about what happened.

When he left from Los Angeles for Memphis, I left to visitMom in New York. I would allow him to get settled first, and then have a couple of rehearsals under his belt before joining. Once I got to New York, I'd be texting and calling him. But, he's up to his normal gig. Out of sight. Out of mind! Finally after Three days go by, he gives me a call late at night. When I reminded him that he was supposed to let me know how things were going, and when I should arrive. He casualy responds with, "Oh don't worry about it. All of my friends from Ailey are here. So, I'm not alone. And, my Mom doesn't think that it's a good idea for you to share my hotel room since the job is paying for it."

Of course! His mother's constant intrusion into our relationship, sends me through the roof. I'm able to surpress my anger at her, to focus upon my anger at him. I'm poor now, but I was never poor before. I had plenty enough money to afford the plane ticket and my own lodgging. But, the bigger issue is that for me to always be there in such extreme manifestation whenever he needs me to be. No matter how big or smal the favor. He consistantly forgets all about me, whenever he has replaced me for the night, or season. And this would be the last straw, and time!

So there you have it. For Four years we did not speak due to matters that trivial and unarticulated. Enough about him for now, by Felisha.

Living on West 137th street was a great way to introduce the concept of independance to my own self. Stripping away all excess, to the point of minimal clothing, posessions, and private living space was surprisingly ideal. A place for everything and everything in its place. The Chi would flow very positively. Providing the perfect atmosphere for creating, exploring , and dreaming. My most favorite thing to do was put on some Arvo Part, gaze outside my window beyond the Urban horrizon, until I reached the place where all I could see was my future. The great thing about hitting rock bottom, is that the sky is the limit for what's to come. My dreams are now so expansive, that they include complete pictures. Complete pictures that run on a time line, like cartoons do. Depicting my every future moments like a film for me to watch.

I naturally fell into this pattern of four hours awake. Four hours asleep. When I wasn't strategizing or budgeting, I was studying or designing. Designing storyboards for possible movie ideas, styling my room. I couldn't do much with the small space. However, I did get Two Chinese Silk Screens that I easily manipulated into about a dozen configurations. Depending on the needs of the day I usually chose the floor plan. And as I got busier, The number of floor plan changes increased. My favorite configuration allowed me to feel like I sat at the front of a spce ship. Only seeing new worlds ahead, as I launched the journey from "Rock Bottom."

As this becomes daily life for me, I begin a new note book. It's a small spiral note book that is sectioned into three colored subjects. I made the Blue Subject, Chore Curriculum. I made the Yellow Subject, Products and Inovations. I made the Green Subject, Log of Evidence.

Determined not to continue in my old school way of conducting business, and getting serious about forming my new company. I pick up an "AM New York." Another great perk to moving to West 137th Street. They have tons of free news magazines for you to pick up right on every corner. In today's I see that a Harlem staple'd electronic store has refurbished computers for only $103.00. So I make this my first Log of Evidence entry. I am going hard at this Law of Attraction stuff, so my evidnce begins with the ask of what I want. Followed by the evidence of when, and how I attract the manifestation into the physical of my asks.

My only income is $188.00 that I receive twice every month from the American Govenment. In addition I get about the same amount once a month in food stamps. So every bill or purchase usually takes more than half of my check. The purchase of the computer, my cell phone bill, and refilling my metrocard took all of my check this period. The thought of going two weeks without any money whatsoever, would have been a deathening thought to the old me. But, I figured I would be so busy creating business plans, broiler plates, company stationary. Oh yea, I've come up with the name for my company. Griffin Cleft.

In my log of evidnece I have a long check list for the type of apartment that I am looking for. Most, of course deal with the needs of my mother. Since we've grown to be best friends, I anticipate her visiting and staying over often. Not having much luck finding a place in this furocious jungle. I give in and hire an Apartment Broker.

I arrive on East 141st street early on a beautiful sunny May morning. I was adoment about not moving to the Bronx. If you watch the local news, as my mother does every night, and never ceases to remind me "Be careful uptown! Another gay boy got bashed or murdered." So my perception of the Bronx was that it was riddled with crime and did not take kindly to homosexuality. I'd never been to the South Bronx before. For someone who has hooked up all over the city, I had never been invited to a hook up in the South Bronx. As if it didn't exist on the New York gay Hook Up Circuit.

My broker arriving on CP Time allowed me to walk around the neighborhood. The first thing to win me over was the number of residential houses in the neighborhood. Growing up in Staten Island, I wasn't use to living amongst concrete skyscrapers that blocked you from looking up at the sky. The schools, and beautifully manicured blocks sealed the deal for me before my broker arrived. Let's call her Ms. Sunday School. As we wait for the Leasing Agent to show us the apartment, Ms. Sunday School tells me about how much she loves the Lord. And, of how active she is in her church. And, that she is also a singer, writer, actress, and anything else you can name in show businees in addition to being a broker.

The apartment is a lot smaller than I would like to have. You know my plan is to split the apratment bewteen living and office space. But, it's a one bed room complete with everything on my check list. Elevator, handicapped bathroom, and even included all utilities. Not only rare for the city, but with a two-year lease, there shouldn't be any reason for denial. As Ms. Sunday School left me to deal with the Leasing Agent, she worries that I wont complete the application. She urges me to go through with it. All that I can think of is how much the application costs. A total of $140.00, and leaving me yet again with no money for another whole two weeks. But, I do complete it. And, on August 1st, I move into my very first New York apartment.

So now my Log of Evidence section is becoming quite full with great findings of how I am turning my life around. Although, at the moment it doesn't feel so good.

I flew accross the country to help my younger sister relocate from New York to Maryland. But, conveniently she choses to be mad at me during my move from Harlem to the South Bronx. So she keeps her car and helping hands in Maryland, while I take trip after trip after trip. Dragging plastic bags and exposed linen down the street, and on the bus, and on the subway. A spectacle in broad day light for all of New York to witness. But, finally it's done and I can begin the design on my new digs.

By this point I miss California so much that I have every intention of doing a grand Hollywood interior. But, it would be a slow build at an income of only $376.00 per month. I mean each can of paint costs $40.00. But, the color palet was decided a long time ago. About two years ago I had this dream.

I was in a theater. And, in this theater an Usher woman dressed all in Black but with a Purple scarf told me that she had all the contacts that I would ever need, and to just follow her. I didn't know how she knew that I was desperate for contacts. And, that all of my LA contacts proved pretty useless once in New York. But, I followed.

Once we arrived to what appeared to be the second floor lobby area of a theater similar to The Warner Theater in our Nation's Capitol. I peek through a black drape that divides the Lobby area from the balcony of the theater. I take note of the red seats and the gold accents all around. As the Usher returns to where I am waiting, I can see a Green EXIT sign just over her shoulder to the right of her. She puts in my hand a black busines card holder that was bound together by rubber bands. Because, the holder was overflowing with business cards, and notes with contact information written in purple ink. "Tell any of them that I sent you, and they should help you."

Of course, I now wake before I can get her name. When I looked down to see the cards not in my hand. I felt the lost of them. The dream had seemed so real, I expected to wake with the contacts in my hand. I thought about that dream for days and weeks. Discussing it with my close friends and family trying to disect the meanings of them. Other than the obvious struggle of trying to create a company from scrtatch and wanting help.

Back to the design of the Griffin Cleft-NYC, yep. That's what I call my new digs. All the while sleeping on the floor. Buying paint to me is actually more of a priority than even an air mattress. Two air mattresses I was given as gifts are now destroyed. I guess their not built for sex. At least, not for how I sex. Especially, after not having my own place for so long. I was over due for some serious fucking.

3 MARK, SET, GO

Today is the day that I had planned to buy my new desk. As I am leaving, and on my way to the train station my favorite piece of music ever written, Beethoven's Moonlight Sonata begins playing in my headphones. I stop and notice that hidden beneath a tree, is the exact first desk I had in Los Angeles, when Jeremiah and I created our Ballet Company. It's a large desk, but it's built on wheels. If I could convince myself to swallow more pride. And, wheel this desk through the busy streets of The South Bronx. I could actually save some money and apply it to the budget of my bed.

Once I wheeled it into it's new living space. With the background of teal green walls with a brown, fiery faux finsh on top of them, my new home slash office began to take shape. The rest should come a little easier now. I just got $5,000.00 in back taxes owed to me.

Yup for the years that I was off, I hadn't filed my taxes. Yes I forgot about all responsibilities and grown man obligations on the way down to "Rock Bottom." Thanks to those SAG residual payments, I still had an income that taxes were being paid upon. So when I finally got to a stable place, I filed. Right a way I stashed $3,000.00 to buy a car for Mom. My White readers who aren't aware, that's the first thing we Blacks do when we get a large sum of money. We run off and buy the best hoopty that our money will get us.

In Mom's case, she has already had a knee replacement and although her health and mobility are improving. She could use a car to help ease some of her pain and stress. The other thing I typically do when I get a large some of money is spread the wealth by hiring someone.

I met AC at what should have been a routine New York hook up. You ever meet someone who you are meant to have sex with? Yet, the conversation takes that awkward right turn towards friendship instead? What do most people do in that situation? I no longer wanted to sleep with him. Yet, I didn't wnat him to leave, embarassed and never wanting to speak to me again. So I threw a party. I told him to invite a couple of dudes over, and so would I.

AC is super model tall, and unfortunately so was all of his friends. How can at my own party, I not have a single soul that I want to fuck. Even the hot guys that I invited to join us were no longer hot. One looked as if he'd just finished eating Thanksgiving dinner, while the left side of the other dude's face looked as though it lost a fight with Tony the Tiger. It wasn't grrrrreat!

The party is a typical Millenial gathering. Everyone on their cell phones texting to other people. Or, checking their Facebook messages. But, AC and I were still having the time of our lives just chopping it up. Just couldn't get horny enough to sleep with one another. By this time in the evening, at least we'd established the feeling was mutual. Still not enough to kick all the horny uglies out. So, I did what any sane Bueen would do. I moved the party to a buddy who didn't feel like traveling. Road trip!

The six of us pile into an elevator down to a taxi, heading even deeper into the Bronx. Praying that we don't get gaybashed or murdered. I send the four of them up when we arrive, but I pull AC back, and the two of us stand in front of my buddy's building and share a cigarette. I don't know what had gotten into me. But, AC and I hailed a cab and bolted back to my place. Who does that? Drops their party guests off to another man's house and doesn't join his own party?

Keep in my mind at my place I still have no furniture. Just my computer and my desk. So there AC and I are, sitting on the floor and tallking. As the sun rises, I learn that both our mothers are from the same town in South Carolina, we both grew up in Richmond, and that we moved to Atlanta at the same time. I didn't tell you about Atlanta, because it was a waste of time for me. But, yea. The similaritiesbetween the Two of us just kept pouring out of AC's mouth.

What you straights might not know about the gay's is that we often have the debate about what makes us gay as well. As our conversation shifted towards that debate. AC introduces a new similarity that is a common core to the Black Gay Male. Many of us will dispute that you are born that way. While I stand firm that it was my choice. One of the laziest excuse a grown ass man can come up with for why they act a certain way, is "My Mother raised me that way."

For most of my formative years, I had my stepfather around. He and my mom split ways when I was about Twelve. He was the type of dad that stayed in the bedroom naked most of the time, so we didn't see him often. Let's call him CK. You'll hear me speak of him often, so remeber that CK equals Stepdad. CK didn't teach me a lot about much, accept for sex. We used to have really adult conversations about sex. Most people would deem them unfit for children today. But in the 80's, people minded their business more. Especially when it comes to raising your own child.

As AC washes up, he explains to me how he had been in a dead end job for the past three years and really wishes he didn't have to go to work. He asked me how he could begin a career as a Fashion Stylist. Do you see how The Law of Attraction works? We'd been up all night sharing the chemistry that would lead us to the entire point of our meeting.

As I go through this period of re-branding and building my family's empire. I am starting by picking up where I left off. The last thing I'd done in Los Angeles before moving was, write a Musical Stage Play, produce Ballet and Opera Concerts, and launched a Fashion Styling House called Ugly Coture Los Angeles. I was commisioning several stage works for Jeremiah to star in.

I had big plans for him to transition form fiercest dancer to house hold name. Kind of like Derek Hough is now doing. So while the writers were busy creating scripts, and I booking his tours. I wanted to create a distraction to Keep Jeremiah busy. The jobs he booked never kept him in Los Angles. So, Ugly Coture Los Angeles was my attempt of keeping an eye on Jeremiah, through styling gigs for shoots and stuff. The firm never got it's official status because, as God would have it. The mouse called him for a tour, and you don't say no to "The Mouse!"

I'd just been reminiscing on old photo shoots and the fun that Jeremiah and I use to have, right when AC asks me for his career advice. Before I could finish explaining to him about test shoots, and mock tear sheets. He pulled his web-site, Facebook page, Instagram page, and EPK for me to peep. I know that I was out of the game for a minute. But, Ac is the first Black Guy I've met on the East Coast that reminded me of how we do it on the

West Coast.

A few things to know about the make up of Los Angeles is that everyone works in the Entertainment Industry, and hardly anyone has children or spouses. In addition to most of us moving from our Home Cities and families to strat new families we cultivate in LA. Working in the industry means that you are always at work. The most casual jaunt into Jamba Juice can have you running into Kevin Hart. Which of course could lead to work, and money. As a result, there is a casual dress, language and attitude adopted by almost everyone. Kind of like there is an LA Citizenship Course taught at the airports.

I notice that AC posessed this dress, language, and attitude naturally. But, he told me he had never been to LA when I asked him if he had aced his CitizenshipCourse. If that's not odd enough, AC also had every marketing tool you could have as a freelance anything. Yet, he stays in a dead end job, that he hates for Three years. It get's even odder when you factor in his quality to detail and understanding the pulse of current fashion trends. Even had shots from a couple of professional gigs that gave him some really awesome tear sheets. The Fashion Industry was as much of a childhood passion to AC, as the Entertainment Industry was to me growing up. You know, that I tested his knowledge of designers. In which, he didn't drop a beat.

"Ok, I'm sold. If you quit your job, I will give you a retainer today to become my personal assistant and Wardrobe Supervisor." It's not just a title. Griffin Cleft will need someone to research and develop costuming long before a performer steps into one. Seemed like a win, win to me.

Here he starts with the resitance that so many who are afraid of success exude. All my life I've been searching for that "Mulan Rouge" type of artist's collective where we struggle our way to the top of historic legacies. But, still after nearly Two years in the city of arts and money, I haven't convinced anyone to join my pod yet. As I looked around my empty apartment with one chair, not even a bed, but beautifully painted walls. I start to understand the risk that I have asked him to take.

Surprisingly enough, he believed in my vision and capability as much as I did his. But, his reservations had more to do with his conversations with God. AC and I shared a common practice that we both have been doing since little boys with no Father. We learn in The Church that we have a Father in Heaven that is always listening, and so we talk to him. In my case, I made him raise me. I always took long walks with God, and still do. As a child our sessions were mostly..... sexual as well come to think of it. But, at any rate, AC had just had a session with God.

As I listen to what he perceived God's message to mean. I had a different opinion. Ever hear the story of the religious man, stranded on his roof in rising waters? God, sends him people in a canoe, boat and helicopter, yet he refused because he was waiting on God to save him. When the man dies, he asks God "Why didn't you come save me?" in which God replies "I sent you a canoe a boat and a helicopter." Yup! Ac, big time. At least now I know how to paint the deal, I go in hard, challenging his faith.

Now he see's me as the helicopter, after missing his boat and canoe. There just isn't a way to be certain. But, I do beleive the epitome of Grand Design has just presented evidence, and pleasure to our Father in Heaven.

I know it could seem selfish for me to shake someone's stability, without ensuring my own first. But, I've been in this position before. The way God always works in my life, is that my wants have to transition somehow into needs, before he shows up. Often when he shows up it seems like he's just in time to take the credit for work I did, and decisions that I made on my own. But, of course in the end when stress and worry turn into calm and ease, we all say, "Amen."

I almost only have the down payment I was saving for my mom's new car. You'd think that she would be doing back flips all the way to the dealership. However, I call her everyday. And, at least twice a week I ask her if she is ready to go look at some cars. There always seem to be an ailment or cloudy weather that prevents her from just spending a couple of hours with me at the dealership to drive off and away from the extra pain and suffering that she is currently embracing.

Each day that mom refuses to join me in looking for a car, I am getting tempted to spend the money. You know for an irresponsible Artist, well capable of some disciplines. The discipline that is Budgeting and Saving is as foreign, as Cantonese is to me. I would give the money to Mom to hold on to, until she's ready. But, Mom is the one who taught me, how to not budget or to not save. So I'd just be paying for her to buy Three Grand worth of Chinese food.

As AC shows up to work more and more, I think it's time I get more furniture. The timing could not have been better. I had my eye on this set from a store right around the corner from me. I was at the store just the other day and saw that the bed, seperate from the set had been knocked down to only Eighty bucks. That along with the Buck Seventy that I saved on the desk, I was rolling in the dough. But, something about seeing it removed from it's intended glory gave me pause. So, after AC took a look at it, he advised me not to. That was just the thing that I really adored about him.

Not, most Black men in our positions would say pass on a bed less that One Hundred Dollars, when the time it would take to actually save up for a bed worthy of our truest character could take an eternity. As we returned to the floor that I was currently sleepin on. I mentioned that I had seen a bedroom set that I would have never thought was my style at all. However, my first impression of it quickly moved me to imagine it in the space. I thought with the grandnest of the wall design. The furniture might need to match. When I set out for a Hollywood design. I never thought the design would be suited for Taylor Swift, or Wendy Williams. But, as over the top as it is. I love it.

I tend to be cheap. We're having the conversation of how the furniture has to speak to the branding of Griffin Cleft. Since, this small space will be shared by personal and career lives. I'm not the best person for accepting market prices for consumer goods. So, When creating my budget, I'd only entered three to five hundred dollars for a bed. In Los Angeles, I have my favorite spots for everything to get at steals. However, I'm not sure there is such a thing in New York. At least not for the customer.

Without even seeing the bed, AC knew that it was the bed for me. I suppose just from witnessing my true dellima of not wanting to let go of so much money on a bedroom set. I wont say how much it cost me. Let's just say that the mattress set alone was more than my total budget. And, letting go of my card when purchasing it was similar to that moment in the movie "Ghost." You remember the part where Whoopi has to give the check to the Nuns, and just can't seem to let it go? It was certainly an act of stepping out on faith. Which is an old idea, all of sudden, creeping back into my belief system. Thanks to AC.

He and I had hours of debates on Christianity, the Bible, and our roles as openly Gay men in the Church. Carefull what you wish for. When I agreed to return the Church, it was only to collect the evidence of where we'd been and connect the dots to where we are currently, and heading to. I made it very clear to my stauchly religious family that I would not be taking Jesus Christ as my Lord and Savior. And, that I would also not be tithing, or taking on the prestegious title of "Christian."

All I'd hoped to gain was conections from fellowshiping with a like-minded congregation, as well as gain knowledge as to how the Earth, Humanity, and the Universe originated. Not sure eactly of what to do with the knowledge. Just, that I heard voices tell me that we all have the responsibility to contribute to the conversations that would shape our world. Scientific exploration has given us great answers to many of the world's most important questions. However, it has not caught up to the influence or contributions of religion. For me, so much is one in the same. The first paragraph of the Bible actually describes what would be the Big Bang Theory. Yup! Right there in Genesis.

When I left the Church, and at least two or three years before I would discover the teachings of Abraham-Hicks. I was almost faithless. Many disappointments of my Church choices really gave me a belief system that compares to Bill Maher. The only reason I wasn't calling Christians dim witted idiots, the way that Bill does. Is simply because, I have seen Christianity work miraculous wonders in other people's lives. Especially the Bishops and Pastors. However, in my own life. No matter how strict I was in my practice, I always felt like Salieri. You remeber how he had so much fond admiration and appreaciation for God. Yet, it was Mozart who alwasy seemd to have Divine Intervention.

But, now after licking my wounds, and arriving back in New York City. I can return to my home church. The Christian Cultural Center is where it all began. And, Pastor A. R. Bernard is an engenious, affectious, compassionate teacher and man. His sermons are never barked at you, "Preaching The Devil Out." As, a lot of Black Ministers do in America. But, instead Sunday service is alot like attending a prestigous Seminary School. Accept a lot more fun, no student loans to pay back, or the committment to putting life on hold while you go grab a piece of meaningless paper.

My philosophy of not giving my heart to Christ is utter bullshit, to AC. He beleives that because I already did as a child, I can't take it back. Hence, Train up a child in the way that he should go, and he will not depart from it. Which was true in a since. All the old lessons and stories were coming back to me. In whole like I had never left. And I promise you that it has been more than Five years since I have even looked at a Bible. Which is the thing that AC is constantly beating into me. Every time I have a question, no matter what. His response will usually be, "You need to read your Bible." This man truly beleived that there wasn't a question you could ask, that a few moments exploring scripture couldn't answer.

All of the other contributions that AC was making, didn't go un noticed. Yet, the main reason I hired him, and the thing that is suppose to become another revenue source for Griffin Cleft, is his passion for being a Fashion Stylist. We're just a few weeks away before we begin production on Mad Science, and I need a new wardrobe for the promotional season.

Not, to mention that most of my phenomenal statement pieces are still in Los Angeles, in Jeremiah's garage. I guess I could have just told Jeremiah to send them to me. That is if he is taking my calls this week.

In television you have the burden of always needing new clothes. Not just to stay trendy,and relevant. But, also because you always need to be in something that you have not warn on a previous episode. When I tell you that I have warn those clothes in LA. I have warn those clothes! So it's time for some new threads. Yet, each day that we set aside to shop for my new wardrobe, he doesn't show up. Of course that doesn't prevent me from buying clothes anyway. As always, when I lay out the outfits for his aproval and compartalization. He says that I picked out everything that he would have told me to pick out.

I'm not sure that is the case, or how true that may be. Since AC has shown up, I have noticed many unatural happenings that are unfamiliar and still positive. The first time that he did not show up to shop, I implemeted a process of only chosing items hangging at the very front of the rack. If I pick it up and it isn't in my size, I put it back and move on. The voices had recently told me, what is for you, Is for you. That bit of guidance makes life a little easier as so many debates regarding the choices of creating a brand, company, and life in a since can be overwhelming.

After a trip to J.C. Penny, AC told me that while I had been shopping, he'd been sleeping and felt that he dreamed of a few of the shirts that I purchased.

Could it be? Could it be that without AC being present physically, he was present and somewhat hands on with the styling of my new wardrobe? Seriously, there was not a single item that wasn't the very first thing I grabbed. No searching endlessly through the ubsurb amount of Larges and Extra Largest to find a Small or Medium, like I would normally.

It wasn't just fondness that made me keep him around. I really felt a since of personal assistance from him even when we were apart. Strangely, during this period, I was working my ass off. Yet still, when I would send AC on an errand, I could not sit that ass down. Even if I wasn't doing anything I would pace back and forth throughout the apartment. As if our movements were being controlled by a single entity that had to make us do the same movements at the same time.

I know how strang that may sound. But, we arrived at a point where we could tell when the other was traveling about. When one of us walked, the other paced around the house, or wherever. At the same time our relationship was entering uncharted territory. We still weren't attracted to one another. His type was "Thugs." Meanwhile, it's rare that I find mysef attracted to someone his height and build. The dancer in me always favors a thick, athletic gym body. So, yes. Football Players all day long.

Somewhat off topic, yet similar wierdness. Every since my nephew Marcus died, my Blackberry has been out of control. When I am trying to move the mouse to a link of my choosing. I have a fierce fight with the cursor, as though the phone would rather I chose differently. I joked at first that I felt that it was Marcus controling my phone. When he was alive, he didn't say much. But, whenever you were looking for your cell phone or the remote control to the TV you could bet that Marcus would have it. Or would have had it, if you'd caught him early enough.

In fact there was only one moment in his entire life where i felt that he spoke to me. But, I wasn't eactly certain because he couldn't muster the control to say a phrase twice. But, I was babysitting him one evening and we were in the basement flipping through the hundreds of channels that you get with cable. As, I flipped past a football game, I heard him say clear as day, "Leave it!" I took a double take, because with all of his unconcious stammering and behaving in typical fashion of a non-verbal child with Autism. I didn't even know that he was present with me as I searched for something for us to watch. "Huh?" I ask him hesitantly, like now I am the crazy one hearing cognisant whole phrases come out of a child that is non-verbal. As, I turn it back to the football game, he smiles, laughs, and then begins running around the room, as to be playing football.

Marcus didn't have a lot of passions he could let us in on. However, Footbal was evidently the big winner. Everyone who knew him knew how much he adored the game. During his final months, he had just joined a football league for children with special needs. His coach was so inspired by him, due to his knowledge and capability of the game being unlike any other child with special needs he'd coached before. More evidence of Grand Design, when you stand at his gravesight, what do you imgine you see in the very short distance behind the cemetary? Fed Ex Field, home of the Washington Redskins.

AC and I are now sharing our crazy. I let him in on the secret that I hear dead people. His response is that he sees them! I hadn't lived at East 141st Street for long, so I could have overlooked how noisy the apartment was. Bewteen AC and I, there was not much left unoticed. As, my attention becomes increasingly distracted by all of the bumps, cracks, and clicks that I hear inside my apartment, I gain the sensibility of us not being alone. Ever! I think most people would have had a similar reaction to mine. I was scared shitless.

There were days where I thought that the spirits in my house was Marcus. While other days I suspect that the spirits of other deaths in the building hung around, and chose to bother me. Strangely, I moved into a building that is prodominantly resident to elderly men and women. Seriously, every month there is a flyer in the lobby for someone else's home going celebration. So, it may not be that odd to assume that someone could have died in my apartment.

I guess the spirits were excited to see AC arrive. The noises in my apratment were becoming so frequent, and loud that you would think you were in some sort of factory. Or as if you were in an old wooden home while outside hurricanes and tornados were abound. Not the case at all. I live in a sturdy brick building, built in the Ninties. Yet here we are with the soundtrack to Dorothy's Kansas. Each time I would here a click, I would jump out of my skin. Meanwhile, AC would be like it's standing right behind you.

Which, he should have kept to himself. Now that they hear someone can see them while someone else can hear them, its like the've just become imortal.

Transitioning to this new company, brand, and chapter in my life was hard enough without spirits or demonds as my sister would call them fucking with me. Not to mention, that Marcus is now fucking with my phone so often that I can't even move my mouse to open my text messages. Forget about replying. If I could control the mouse long enough to make a call, that's how I would reply to everyone.

I have always been some what aware of communicating to the "Other Side." Typically, sleeping is where I did most of my communicating. Often my dreams were quiet reflections from my life, or how I would have preferred my life to be. But, there are also moments when dreams provide direct access to someone who has died, and want to communicate something back to earth.

My Grandma Molly passed away, leaving her Million dollar historic home to my cousin who was not prepared to deal with owning such real estate.

That's in addition to mourning the passing of his Grandmother. Not to mention, his mother pasing a decade before Grandma Molly's passing.

So as I visit the mansion for the first time since Grandma Molly's passing, I notice that the home is in the poorest condition I'd ever seen it in. The animals were unkept or fed. The Gas Bill had become dilenquent and shut off, so there was not even hot water or cooking gas. And, my cousin didn't even leave his bed, but to use the bathroom. I'm talking Robert Zemeckis pitiful. I couldn't tell if the cats were following me around the house because they needed, food, TLC, or just excited to be around someone with live energy.

I'm willing to bet anything that it wasn't any of the above. I've dtermined that they actually had been following Grandma molly's spirit, who was clearly trying to communicate something to me. That night she came to me. In a dream. I can't remeber what our conversation in the dream was regarding. However, I got the gist when I woke up balling. I have never cried this aggressively before in my life.

Cries so deep that I couldn't catch my breath. When I could muster enough gasps to formulate a word, all I could get out is "I can't tell him." I rolled back and forth crying and repeating "I can't tell him!" I'm not sure if those were my own tears and words, or Grandma Molly's. One could assume that they weren't my own, since I woke from a dead sleep. Also. What could Grandma Molly tell me that I couldn't tell my cousin? If that truly is the magnitude of what we are dealing with, I'm thankful that I don't remember it.

As you can imagine I am way over due for a trip to Los Angeles by now. I have a very dear friend of mine who throws a fantastic Annual Gala for her opera company in celebration of Juneteenth. It's approaching faster than AC and I can drum up new revenue streams. Though, I am optimistic about the relationship that is blossoming with AC. Ideally I plan to take him with me to Los Angeles, which will be his first time.

I call Mrs. Four Octives my "play, play sister." We met Twelve or Thirteen years ago, when she hired me to dance for her Opera Company. Opera Noir is a West Coast Jewel. It's founder, my play play sister created it to promote cultural diversity among the classical arts. Through the years we've known one another, she has produced and stared in numerous Opera Concerts where she has provided a grand forum for Blacks and Latinos to experience the extravaganza that is The Opera.

I fell in love with the son of Mrs. Four Octives first before adopting her entire family as my new West Coast peeps. On my first visit to the Four Octive Home, I noticed a wall of wonderful photography of her son. I didn't understand why looking at this little boy in the photos would make me tear up. When she noticed my tears, she explained that her son was diagnosed with Autism. Meeting Kent, her son, was my introduction to the world of families with children of Special Needs, hence the tears.

Kent did not posess physical diformaties in his face. Like, the children with Down Syndrome. He appeared perfectly normal in his photos, yet. I could read the energy from his eyes telling me very clearly, of his unusual story. As the relationships between the Four Octive family and I grew closer and stronger. I grew from family friend, to baby sitter, to Uncle Mark.

When my biological family received the news that Marcus had been born with Autism, like lightning jolting a direct bolt to my heart. I knew the road my oldest Sister was destined for. The struggle. The embarassment. The guilt. The constant questioning of God's purpose for her son. But, I could not predict that at Nine years old, he would die. With Marcus gone I was especially excited to spend some quality time with Kent when I returned to Los Angeles for the Juneteenth Gala.

With my Summer/Fall wardrobe pretty much styled and purchased I shift AC's attention to Griffin Cleft's Mad Science. My idea for the web-based docu reality show was to take the printed pages of a production log, and give it legs on film for audiences at home could follow the development of Griffin Cleft-NYC. Bearing witness to the process that God was implementing in my professional life.

I had already begun noticing the changes in energy of my spiritual world. Although, I was so removed from the Church, and for many years now. I had never felt closer to God. Finally, it appeared as though the spiritual equity that I had invested in years of being a faithful Christian was garnering tangible success. By tangible, I am speaking in reference to communication.

It's unexplainable how my conversations with God as a child gave me solice and comforting answers. But, now as an adult I begin to hear God's voice very clearly. His voice is deep. Extreamly propper, yet posesses an exciting lilt similar to a game show prize announcer. But, with a much more charming Swagger. Dare I venture to say sexy? He definitely sounds like a DIlf to me.

Until now, from the time that I'd lived on West 137th street I had been hearing the chant, "You can't get it wrong. You can't get it right." It had grown to become a rhythm. A pulse like a heart beat that drove the timming of all my actions. Over time the single female voice was joined by more and more voices until finally it sounded like a chorus of Thousands surrounding me as I made every move, and hammering this concept of me, neither getting it wrong nor right.

Presenting the idea that I could communicate with the other side on Mad Science would prove to be harder than I anticpated. When I would watch the dailies back, the energy of real life moments that seemed to fill my home was absent. I appeared so lonely, even though I could feel the energy of Thousands walking with and living with me. Not to mention that I could also hear them.

So as AC and I explored ways of capturing the truthof what was manifesting in my life, of course we'd typically end up at his go to answer, "You need to read your Bible." I had only owned a New Testament Bible at that time. And, I also didn't know if I was ready to go back to a doctrined life. The time has come for us to have the conversation. AC could not understand why someone who was so connected with God. Someone who could repeat scripture, had loads of faith, and understood his God to be the father of Jesus Christ would not want to give his heart to Lord and profess Jesus as his Lord in Savior.

I was determined to learn everything there was to know about Christianity before I gave my heart and soul to the Lord. After discovering Abraham-Hicks, I adopted the belief that even after death we are giving free will and choices. If this true, living eternally in a place called Heaven. Where all is well, and boring, is not attractive to me.

So I thought I should be someone who has practiced the faith, and studied it. Yet, I did not want to be a true Christian destined for a boring afterlife in a place called Heaven. Especially when I was clear that the answer to Earth's population problem is intergalactic travel. Though I'm aware that it wont happen in this lifetime. I want to keep coming back to earth until it does. I want to be one of the first Humans to live life on Mars. As well as take a leading role in navigating the journey of other Earth Citzens to Mars.

Martian Life may sound crazy to some of you. But, the NASA space program played a recurring role in my upbringing. One of my first gigs after arriving in Los Angeles, was an industrial video for the NASA Mars 2010 program. A gig that Jeremiah and I both were hired for. Here was a program designed to transport the first group of Human Civilians to Mars. While on the job, I learned that the Civilians selected would travel in a confined space shuttle for Two years before arriving on the planet, and would have to live the rest of their lives in this expansive dome like space center that would assimilate the characteristics of Earth. But, they would never be able to return to Earth because of the changes their bodies would inccu, in terms of gravity, weight, and such.

I was happy to contribute to NASA's initial attempt. But, I was clear that this would not be the lifetime where I would join them.

As I begin to move more furniture into Griffin Cleft-NYC, the space is really taking on a new life. Not being able to see the faux walls, in completion anymore allows your focus to connect new shapes. Kicking back and sharing my story of Intergalactic aspirations with AC, when all of a sudden it hits us. AC poins out that I had painted my story on my walls.

My intention was to scallop the borders where the walls meet the ceiling, giving the perception of a Morrocan-like, Castle. Kind of like Princess Jasmine's quarters. Until AC mentioned it, I had not known that I had painted space shuttles. On my East Wall I painted Eight of the thinner old school shuttles. And, on the West Wall, I painted Four of the thicker newest models from NASA.

To answer AC's question, becoming a true Christian wouldn't allow me to return to Earth anymore. So I figured that I'd enjoy the study and practices of Christianity again, and again, and again, and again, then finally. Say, ok! I've had enough. I can retire to Heaven now and live eternally with Jesus Christ. In his mind I had it backwards. "It's not like you have studied Theology at the best Seminary in the country. How can you expect to Understand all of Christianity, when people who have, don't even understand it all?"

Wow! Ok, Ten points for Mr. AC. And, just as I was about to kick him out for the day. He reminded me that There is so much that I was taught in Sunday School, as a little boy. That I now have forgotten, that will resurface. Train up a child in the way that he should go and he will not depart from it. I did notice through our debates, I could dispute and present religious facts in arguments. I guess he's kind of right. And, I suppose if I didn't enjoy the true Christian experience, I could always walk away from it again. "What do I have to do to make Jesus my Lord and Savior?" I blurt out abruptly.

"Repeat after me" he says. So I do.

4 STANDING WITH TRUTH

My family will be so excited to learn that I have returned to our faith. I'll have to chose a new home church. I'd love to join CCC, and continue being led by Pastor Bernard. However, with the Subway's Sunday Schedule. It will take Two hours in both directions. The last time I was a Christian, I was extra. This time, I have to make it as coveniently, comfortable as possible. I also hope that Jesus knows that I will be going back to my old tithing habit as well. I give my tithes and offering to my neighbors, and community organizations. Not the church. This way I don't anticipate God giving me back my money Ten fold. I hope I can still be Christian with these conditions. I'm too grown to adopt every law and commandmant of Biblical times to my Modern Times.

All and all, I feel it's the right choice to make. The biggest of my issues with Christianity before. Is that I was in search of a belief system that worked One Hundred percent of the time. After discovering the Law of attraction, I have learned how much is at play in the efforts to manifest success. Even if it were as simple as praying for it. God would have to find a way to communicate, distribute, and invoice you for the delivery of your success. I'm finding that the easiest way to look at beliefs, is to see them as tools. Then, I identify which tool to use for each individual repair.

It's been about half a decade since I have used the left side of my brain as much as I have lately. Debating Christianity, and life with AC exausts me in ways that I have never been before. I've been procrastinating for the past hour. My intention is to get ready for bed. But, I seem to be stuck in my desk chair. I don't have the energy to turn on a light or even the TV. And, I've been sitting in total darkness.

I'm not sure that I was thinking of anything in particular when I began to hear sounds of static electricity charging. I looked around to see where it was coming from before realizing that it was coming from my body. Out of no where my body begins snapping and popping. You know how you get a jolting shock from dragging your feet along carpet. It was like that except, a lot of them all at once. I would guess about Twenty Five or so.

They were coming from my back, so I couldn't tell if the electricity was exiting or entering my body. But, I could feel that something unusual was happening.

I didn't feel any less normal, when my electicusion was complete. It's almost as if energy had entered my body. Performed an assesment of soul and/or heart. And, then left the building. Hearing of it many times in Church, I wondered if it was the Holy Ghost. Everyone always speaks of the Holy Spirit at work. But, not much description about how its presence effects the Natural. Ten more minutes go by. Still I sit there in darkness. My first thought is to wonder if I've caught the Holy Spirit. Will I speak in Tongues? I've always wanted to have that experience of falling out in Church, speking in Tongues, while lifting my tear drenched eyes to the sky in praise.

All of sudden I hear in a Woman's voice. Very soft spoken.

"I am the Spirit of Truth. I have been sent to you, by God, to let you know that God is calling upon you. You have been chosen to validate the presidency of Barack Obama. This has always been your purpose for returning to Earth. To become one of God's represntatives during the Christian Validation of the Forty Fourth Presidency of the United States of America. The Christian Validation has already begun. Are you Gay?" I reply with, "As a Broadway Musical." I took her silence to mean, now wasn't the time fr joking. "Yes. I am Gay." "Was it of your choice to be Gay?"

Not sure if now is the time to change my position on whether we are born with sexual preference or not. I thought, that one day, I would marry a Woman. Create a family, and have an understanding with my wife that I would still have sex with men on the side. But, since that hasn't happen yet. And, I've grown so accustomed to my freedom, that marriage has become farther and farther out of sight to me. So, "Yes. I choose to be Gay." I say to answer the Spirit of truth's question.

"Don't change your mind. You must have the courage to stand in the truths of your decisions. Remember that abandoning your truths, abandons your purpose." She pauses. And, everytime she does, it's as if she leaves the room. I can totally feel her presense when she speaks. But, not at all in her silence. "Are you still here?" I ask. "Yes, I am still here." She replies. "Okay....?" I whisper. "You have Two important roles to play in moving forward with the Christian Validation of Obama's Presidency. The Gays of America will decide the Forty Fifth Presidential Election. You must become the first True Christian Man, to Marry a Man. And, you must help to make the Presidential Race, for the orty Fifth Office, a fair Race. Do you understand and accept your calling?

As she explained that to me. The chant that once sounded like the furocious roar of an NFL arena crowd begins again. But, this time there is only one voice. "You can't get it wrong. You can't get it right." This is when I first hear the voice of my sexy prize announcer, God continues...

"This is not about sex. This is about World Peace. We have tried this many times before, and failed. I will place upon you a dual task. It will begin with your Christian obligations to the United States of America. Afterwards I call upon you to aid me in a far greater task. The task of creating the New World."

Do you see what I am talking about! I'm a big dreamer, because nothing I've done ever seems big enough to match the purpose, that my inner being was certain would come one day. All of my life. No matter how ghetto, or discouraging my situation had become, I felt deep down inside, that I was meant to accomplish big deals in my life. And, God himself has come down to get me back in line with the core of who I am. And, what purpose I was sent here for.

God continues. "No one ever dies. You simply transition worlds. You can't get it wrong, and you can't get it right. If you meet my hope for you, the World will have peace. If my hope for you exceeds your capability, you will have a successful company. If there is such a place called Heaven. You are now in it. Where you are is quite new, and so are you. In fact it is all forming as I speak. Your transition begins now. It's time to make the journey to "Mark's World." You wont be able to find your keys, so leave your door unlocked and meet me down stairs."

This is the voice of God you guys. Am I supposed to really look for my keys to prove him wrong? No. I bolted outta there faster than you could say Jackie Joyner Who?

"Walk with me." I am not sure of how I knew which way to walk. But, it was though my body had a GPS system that took over. My words, actions, and thoughts no longer seemed like they were my own. I didn't know if I were being controlled, or if I had died. So many questions were going through my head. Mainly, how is it that I can hear a man calling himself God as clearly as if he were standing directly above me. Yet, I couldn't see anyone else around. Like I was truly on an Earth, where I was the only living creature. No neighbors. No shops open. In New York? The Bronx? If you are familiar with the city, you understand my concern.

He asks me "Can you remember everything you have done in your life, and press a button within five seonds?" Not understanding the magnitude of the question, of course I answer yes. I figured if God chose me, he didn't want to hear no. He proceeds, "Run to the stop sign and back." I run and walk back to him. "Remember my words. I did not say to walk back. Now run to the stop sign and run back to me." I do. "Now run to the stop sign and run back to me. When you get back to me fall to the ground." I do. "Now run to the stop sign and run back to me. When you get back to me fall to the ground, roll around, stand up and scream at the top of your lungs while jumping up and down.´ And, I do again.

By now, my body has figured that God is testing my memory. Like I had been pre-conditioned before birth to expect these tasks to validate who I am. So I am getting into it! Not only am I lazor focused. But, I am having all the fun a todller would have, playing with his Father. Did I mention it has begun to rain since I came outside?

"Now run to the stop sign and run back to me. When you get back to me fall to the ground, roll around, stand up and scream at the top of your lungs while jumping up and down. Then, throw yourself in that pile of garbage bags." I do. Effortlessly! "Now run to the stop sign and run back to me. When you get back to me fall to the ground, roll around, stand up and scream at the top of your lungs while jumping up and down. Then, throw yourself in that pile of garbage bags, get up and throw your keys as far as you can." Where'd my keys come from? Were they in my pocket the whole time. "Now, retrace your every step backwards and go find your keys."

The lesson, I guess, could have been that I could have retraced my steps only once. But, in that mindset, thinking child-like. I retrograded the restart of each starting point of the memory game. Which took longer than we thought it would, and by the time I looked for my keys, I was looking for them in the dark. I did not find the keys, but luckily I left the apartment door unlocked. As I returned to the apartment I felt played. Like God's sole intent of the exercise was to have me loose my keys.

The next day I woke up feeling as though it had all been a dream. I could no longer hear God's voice as clearly as I heard it the night before. It was a Sunday morning. I didn't have a lot planned, so I would take the day easily. It was a beautiful morning and afternoon. I thought about God's and the Spirit of Truth's visit all day. Oddly enough, I could not remember what I dreamed last night. Odd, because I always remeber my dreams. Was my memory broken? Did he take it with him?

AC and I met up that day to buy a TV for Griffin Cleft-NYC. Since it was his job as stylist, to style the fashion of the office. I thought he should be included in all big purchase as well as office designs. As we visited a neigborhood electronic store, they convinced me to come in and apply for credit. I knew my credit was shit. But, I also knew that I had a shit load of money deposited recently from those back taxes. So I went for it and applied. To my surprised they qualifed me for a line of credit. Thinking that it would be a good way to begin repairing my credit, as well as hearing God's voice, just briefly to tell me to buy the television. I buy a Fifty inch flat screen that was double the price of what I would pay at Walmart.

After we get back to my place. We begin to fellowship, like we always do. He asks me where my resistance to excepting Jesus as my Lord and Saviour came from. I expressed to him my story of wanting to be on the leading edge of Human's transition to Mars. But, I also shared with him what I knew of the creation of the Bible. All of the books that originally were extracted, by Helena of Troy's Holy Crusade. Which, she basically compiled Isreali children stories she had been told along with divine intervention she claimed to have experienced during her voyage. And, the major piece was that the phillosophy taught by Abraham-hicks, which was working so well for me, did not include evil and bad. So, it was hard for me to beleive in the Devil. Or, a doctrine that included him so prominently.

Just like it didn't take him long to convince me of the power of Jesus Christ. It didn't take AC long to show me the power of the Devil. As he began to tell me stories of him witnessing the Devil at work. I felt the presence of the Devil. Like he'd shown up to gloat in the tells of his horrific works. This, dark and strange energy was very scary for me. We immediately began praying. "I don't know how to make him leave." I yell out. "Recite the Lords Prayer." He yells back at me. "Our father. Which art in Heaven....." we recite together. "He left." I say. I could feel his exit. "But, I bet he's still in the building." AC, exclaims.

Totally in sync. The two of us, run into the hall. Banging on my neighbor's doors while screaming "Get the Devil out!" Could you imagine? Two of the queerest looking Black dudes, bombarding your door like the police. Yet, instead of screaming fire. They scream "Get the Devil out!" When we weren't screaming that, we were screaming The Lords Prayer. I even called up Mrs. Four Octives who I knew was believer, and asked her to pray with us as we got the Devil out of my building.

You would think that was enough to get us off the topic of the Devil. But, nope. As soon as we get back to the apratment, him too I suppose. Our conversation moves to an Ugly debate of whether or not the Devil is of God. I argued for God creating the Devil, and he argued against it.

When we did our research he could not dispute his argument. Where as I could find much evidence that supported my argument. In through my quest for evidence, I could find that God created him to be the adversity that would educate and challeng his son.

Through our fellowship. AC and I grew closer and closer. Having so much and common, I began to feel as though he were my soul mate. I was in a period of studying everything I could lay eyes upon. So I looked up the definition for soul mate. Definded by Webster. A **soul mate** is a person with whom one has a feeling of deep or natural affinity. This may involve similarity, love, romance, friendship, comfort, intimacy, sexuality, sexual activity, spirituality, or compatibility and trust.

Most was present except, the sexuality piece. I was not attracted to AC in the least and neither was he to I. Just as I finish reading the definition from Webster's, God shows up again. He starts by teling me what a Soul mate truly is. He gives me the analagy of polar opposites. But, defined very differently than we understand Polar to be.

He splits the soul into Two parts. One stays in the natural world. While the other stays in the Physical world. And, like clock work we trade places. The image he often gives me is the shadow of me walking on Earth, while connected to my feet is another me, taking every foot step the way I do. As if I were walking on water and could see my own reflection mirrored through Earth. However, he made it very clear that the reflective perception wasn't an accurate one. Not only are the mates their own respective entities, living and fuctioning both at once. But, in separate worlds, with contrasting purposes. In fact, he has every characteristic I lack. Every ability I have, he lacks. We're polar opposites. However, when we switch sides. We both take on similar Earthly roles, and similar Spiritual roles. But, we don't inhabbit them within respective accordances.

He also pointed out that sometimes your Soul mate can disonnect and make his way to the Natural World while you are still there. That is what has happened. "Can't you tell from the way he describes your mother?" God asks. "Wait? You mean to tell.." I begin to ask, before he interupts me with, "What are the odds of Two Sarahs from Greenvile, South Carolina producing Two Black Men that only differ in height?" And, God was right. AC and I were so similar, it's as if he watched my life on a TV screen and regurgitated to me.

My debate with AC now shift gears from the Devil to how muc we were reminding each other of, ourselves. It didn't matter that I hadn't been to Church in years. That saying "Train up a child in the way that he should go, and he will not depart from it" definitely rained true here in my life. Suddenly, I notice the super natural. I could feel the energy of being joined by others. As though the spirits of the Bible were all coming to gloat in their contributions to the stories that created Earth. And, how new generations of Americans still believe, debate, and respect them.

AC even picked up on it. Like I told you before, AC was just as spiritual as he was religous. Of course he would be, because he was me. I could hear what he could see. He could feel what I could sense. As I got clarity, without discussing such clarity, so did he. But, I must have gotten the part of our brain that proccessed clarity. After a day would past, he could not remember any of it. It came to a point where our debates mostly consisted of arguments that we'd had in the past days or weeks.

Just as we get into another one of our debates. Again, about the Devil my door bell rings. It is the company I called to install my entertainment center that God made me buy. When I open the door, this Black man about Six Feet, Two Hunderd and Fifty pounds, with long dred locks, yellow eyes, and holding a hammer, stands at my threshold.

The White girl in me, jumped out screamed and slammed the door. I ran back to AC and yelled "It's the Devil!" Franticaly, he grabs my hand and we both start reciting The Lords Prayer. As we do, I hear the hammer hit the floor outside my apartment. So relieved, as I took that as God's signal that we were safe. So I went back to the door to let the guy in. As he came in, a tiny woman unfolds from behind him. As I apologised to them both. He explains that the woman was his girlfriend who tags along with him, watching him work. And to keep him company, I guess.

AC and I get a glimpse into what our relationship could grow to. They ask us if we were lovers. We both say "No, Soul mates" They're looks of awh, quickly turned into dazes of confusion as we explained our definition of Soul Mate, and how our relationship could appear to be a marriage. But, it's moreso a Life Arrangement, where we assist one another for the purpose of creation. Introducing this concept for the first time on fresh ears allowed us to see what people's reaction would be to our depiction.

"There are Three things that God wants from a Man." This is the pastor from my home church speaking. Although it's as though God is speaking through him directly to me. Now that the nice couple has installed my Entertainment System, I can watch T.V., finally. Not just T.V., but with the new Smart technology I can watch YouTube right there on my set. Being absent for a while, I'm anxious to find out what my Religous leaders are talking about. So, I look to see if there are any videos of Pastor Bernard. To my surprise, there were many. He continues, "The three things are,

Decisiveness, Consistancy, and Strength."

I get it. I have been very decisive in what I want, need from and plan to do with God. I have not been so consistant. In fact I have been all over the place lately. But, the strength piece, puzzles me. Does a strong man fight to the death, or does a strong man take the high road, not fighting. God quickly disrupts my debate with an answer. "Your purpose is so important, that you must not fight. In your ghetto, you will be the odd ball that will run to the police. It is what they are there for, and why you pay their salaries. They are there to serve and protect you. And, you must be a law abiding citizen. We must keep you alive, out of prison, and healthy."

God continues, "But, you musn't fear death. For, if you die you'll only awake next to me. The only thing you must fear is the Lord." Noticing the oddness in him refering to the Lord as if The Lord was something removed from himself. I quickly begin researching The Lord. There is not much clarity on the subject. Since many theologist have more spiritual understandings of The Lord than a natural depiction. Finding an exact science in Spirituality is a futile task.

When AC comes in for work, I share my news of God's visit with me and Pastor Bernard. It doesn't take long before we're at it again. Me reminding him of the position he took in yesterday's argument. Enough is enough! "Let's do what God has been trying to get me to do for weeks now. Let's turn to his Law on Earth, the Holy Bible." We started from the very begining. Until now I have faught reading the word of God, because I just wanted him to repeat everything in my ear.

I had AC read, "In the begining, God seperated the heavens from the Earth." There are freudian slips. And, there are monumental fuck ups, like misreading written text, that's right there in front of you.

We spent the bulk of the day constructing the optical theory based on the text that would conclude findings similar to the Big Bang Theory. After all the debating, and I read what the opening really says. I fall into a sever state of doubt. "It's all bullshit!" I ye'll out for the first of many times.

I will save you some of the profanity, just know through all of my studying during the Purple Period, I say that a lot. Usually because I would dispute scientific and religous facts better than anyone I had heard before. Me, the dude without even a High School Diploma. Only to find that my arguments had no foundation, due to him mis reading, or misinterpreting passages. Sometimes even misreading Wikipedia definitions of words crucial to our debates and findings.

Why did I continue to let him read? We were using multiple cell phones, tablets, and computers as well as written resources. I couldn't do all the reading by myself. Also, his devices would use different search engines that would yield different results. Now it has become as though instead of building an Entertainment Company, Griffin Cleft-NYC, is becoming the base for Christian Theology and Validation. Lead by God. I say lead rather than taught, because it seemed that if we couldn't make our way to the truth on our own, it did not or would no longer exist.

Whenever I read the Bible. I am like a Four Year old asking their parent, "But why?" Over and over again. I often read with conflicting understanding. Of course, subtracting the antiquated comparrisons from the text that is no longer substainable, in today's Modern Times. The start of Genesis is the perfect example.

The first line correctly reads,"In the beginning, God created the Heavens and Earth." That in itself is not a propper or clearly articulated statement. In the beggining of what? Earth? All time? The Universe? Christianity? If he created the Heavens and Earth at the same time, does that mean that Heaven and Earth are the same age? What was the purpose of Heaven, if no one had died yet? And, if there are multiple Heavens, when was it decided upon a singular Heaven where we live with God?

Excuse all my questioning, but, it's hard for me to accept things just because people tell me to. We got to break this shit down.

If you read further down to the end of vers two. You'll read that the Spirit of God hovered over the water. The Scientist in me knows, that. In forming the Earth, Fire came first and water second. How much time passes between vers one and vers two? Or, is the starting point for the bible after the big bang? Which would then make "In the begining" pertain to a marking point sometime after years offire burned, creating the foundations for the water to lay upon?

"They want to tell you something, but you have to find it for yourself." AC tells me out of the blue. "Find it where?" I ask. "Read your Bible!" He ye'lls out. I quickly close the Bible and toss it. Have you ever seen that thing? But, AC wants me to read more than Fifteen Hundred pages? Now?

He refuses to speak at that point. Or, as he tells me, "They wont let me tell you anything." I'm asking him questions, and he keeps shaking his head, no. So, I pick up my Bible and I begin reading. As I skim the pages, I am clueless as to where to start. "Can't I have a Clue?" I ask.

Then he begins to give me clues by using his body and pointing at various things in the house. He'd point to a door nob. I would notice the metal of the nob, and yell out "Gold?" He would then catch the Holy Spirit, go into body convulsions, and then fall to the floor. Like he had just witnessed the Sermon on the Mound himself. So I'd quickly begin searching Google for the origin of Gold. Learning everything about it from it's use as currency, sports club deligation, to it's uses by God and the Bible.

This went on and on. His clues were as vague as colors, clouds, and female. I had no clue what I was actually searching for. By using my tablet, smart phone, computer, and Bible. Even a book of prophecies from Nostradamus. I didn't know what made him stop, but finally.

"Yes." He barely can get out. I mean he went through a major physical ordeal just now. Almost as if his body just went through hours of seizing. Beter yet, an orgasm for hours is what it appeared to be. He continues to speak as I look at him like he's crazy. "They would not let me stop you. It started out as a simple test to see if I could motivate you. But, you surprised them with your committment." "Who the fuck is they?" I yell in confusion. "God is not by himself." He yells back.

I'm not sure whether to be envious that he can see God and I can't, or if I should be running for the hills because this motherfucker is crazy! Just before I was about to tell him to get the fuck out. "They're amazed by you." He tells me. "You just read the Bible in Five hours." AC explains.

Yup! Five hours of running back and forth between tablets, books, Bibles, smart phones and computers. Five hours of watching this Six feet, Six Inch man roll around on my hardwood floors, orgasing and foaming at the mouth. Five hours of heavy breathing, moaning and sighing and, marveling at how his clues led me to the next History lesson that would end in my knowing the entire story of Human Civilization. That's after only starting to read it Five hours ago. It was like the end of that film "Lucy." Accept what Lucy saw in Five minutes, took me Five hours to see.

"Your number is now Five." I had no clue what that meant. I figured it had to do something with how long it took me to read the Bible. But, there was no time to ponder since my training continues, as soon as I wake.

"Safety first. You have to start prepping your place for children." This is God still talking to me. "No longer can you leave lit candles burning when you leave the house. Always unplug the stove when you are not using it. Harmful chemicals must be moved from the lower cabinets and moved to higher, cool and dry places." And, so on, and so on. The orders come as quickly as I can move. "Time to go." He says. "Ok, let me just shower first." I tell him. "No time. When I say it's time to go, you must leave within seconds of me telling you. Not minutes." He say's, while almost pushing me out of the door.

I'm clear that there is no physical feeling of touch from God. But, I'm telling you. Our bodies must come equiped, with the knowing that the moment could arise when you are chosing. And, it knows exactly what to do. Not knowing where we are going, my legs walk without confusion. We end up in St. Mary's Park. At this moment I feel my mind check in with the Sun. Weird that I could tell. Yet, it seemed as though my GPS was accessed and control by the Sun which happened to be to my left at a direct Right Angle. Now a voice takes over that is not God. Unless he changes his voice. But, now the voice sounds like a Red Neck, military Jug Head.

"You understand that you come from a military family?" I'm asked. This is definitely true. I'm the only one of the men in my family that chose Ballet over a career in the US Military. My Father was in the Navy. My Brother was in the Marines, while my uncle was in the Air Force. And, Mom's first husband was in the Army.

The Military voice continues. "Make a left. Up the hill." I do. "Good, now as you walk down. Take your hat and ear phones off." I take off a Russian Lumber Jack hat. You know the one with the floppy ears? And I fold my ear phones into the hat. "Now as you walk by the trash can drop them in." Are you fucking kidding me? I love these ear phones. It's my best fitting pair to workout in. But, I still drop everything into the trash can as I pass it.

"You see that line of children?" I'm asked. "Yes." I reply. "That number of children will fit into that White van you're standing in front of." After hearing that, I'm a little creeped out. Does he expect me to kidnap some children. Meanwhile I'm still walking in zig zags like I just debarked some Cruise Ship.

My body then takes me into a discount store. I'm led through out the store pilling lots of outdoors type of products into my shopping cart. Games, Inflatable swimming tubes, Beach balls and umbrellas, ice coolers, ice packs, cups, first aid kits and products, oversized pillows and more. You get the gist.

"You're part of the disaster team. You will keep the children. You must keep them occupied, hydrated, kept cool, and most of all. Keep them from looking outside." The bit about keeping cool leads me to believe that this has something to do with Global Warming. Oh yes, I may have forgotten to mention that, I ave no idea what the hell I've been chosen for. I know what I heard. It's hard for me to believe any of it, since my Faith has been so damaged. I've yet to regain the trust for God that I once had. Although he reminds me constantly that he was there, right by my side when I left the Church. I unfortunately forgot all about him.

As I stand in the check out line with my cart full of things that I would not buy for myself. Some how, I'm suppose to pay for all this shit! Am I going crazy? As I total the cost of the items, and get ready to walk out leaving the cart and everything in it right there with the cashier. I start getting text messages. The messages all include media files. As I flip through the pictures that start out pitch black. They get lighter and lighter until finally I see a young boy sitting at a small desk with his head burried in a book.

Is that young boy me? I wonder. "Yes." I hear God say. "Where were you just now?" I ask in confusion. "I've been here the whole time. Nothing you do, say, or think escapes my witnessing of." Before I can wonder off into the emberassment of God being there for some of my most degrading moments, I flip to the next picture. Which has now brightened to clear as day. I can see now that the book that I am studying as a young boy, was the Psalty's Kids Bible. My first Bible, and I stil have it today. Standing above and behind me to my left and to my right are Ailiens.

I can tell you from studies that move beyond this Purple Period, that God does very little to disrupt status quo. So, if he shows you an Angel or an Ailien. They would appear as you expect them to look, and not actually what they may look like. In this case. Standing above a young Black boy. A young Black Boy studying the Holy Bible. Are two clones of "Paul, the Ailien." Accept they were much taller. As if Paul was a child of the present two adult Ailiens. Long, thin grey legs. Gold waists. White "V necked" T-shirts. Long Grey necks that widened into Grey egg shaped heads. Big, petruding, Black diamond shaped eyes.

That should have scared the living day lights out of me. Yet, it were as if my body had been expecting this confirmation for a long time. As though I was finally getting validation for things I'd suspected for most of my life. But, to finish the mission at hand. God and I had to discuss who the hell was paying for all this stuff in my shopping cart.

"That's all for today." He tells me. "What do you mean?" I ask. "The purpose was not for you to purchase those items. It was so that you would know what to purchase when the time comes for it. Your obedience is to be rewarded. Now go outside and see what you've won."

As I leave the discount store. Nothing seems to be out of the ordinary. I see a moving truck unloading furniture. So I run to it alerting the movers that I was the prize winner. They didn't know what I was talking about. So I stand in front of the store for a while awaiting and looking for my prize. After about twenty minutes, I start heading home. I make a right at the corner and see a short handicapped school bus parked with kids lined up on it, and ready to exit off of it. Since my activities today have all been around safety, and children. I run over to the bus, with the knowing that these kids must be waiting on me. As I get closer to the bus.

"No way. Could that be?" I wonder aloud. I can't make out his face, but certainly from the back he appears to be my little nephew, Marcus who passed away Two years ago. Most of you are thinking, this dude must be going crazy. I'm telling you that my expectations of God's power were wildly imaginative. There wasn't a thing that was impossible for God to do during the Purple Period. Or, so I thought.

As the last child exited the bus. The one that I was sure would be Marcus. And, the bus for special needs pulled away. I stood there alone. Without Marcus, and without any reward. Nor a prize. I headed home. But, not before thinking to myself, "This is bullshit," yet again.

5 GOOD HELP IS HARD TO FIND

I have gained new perspective on the term, God's Children. If you are ever so fortunate to be chosen. The sooner you choose, the sooner you'll get your chance at being chosing. Of us chosen, we get how playful the spirits, and Gods are. They would love for as many of you as possible to be chosen. Then you too, will understand how playful the spirits and Gods can be. On the other side it seems as though everything is just fun and games. Well, in a way. The work is real, and constant. Yet, they seem to have very intersting ways of conducting the work.

God among others, already can tell when you haven't arrived at the fullest understanding of purpose, plan, or existence, for that matter. It seems that every second of the structured process in which we conduct our work, is only leading me twoards more complete understanding of what it is that we are really doing. Such is the exact manipulation of guidance.

But, it is my hope that they can be more careful, in the future. The best way to decieve someone, is to make them think that they are enjoying the lie. Especially since this gift has come at such a desperate time in my life. As I watch all of my friends on television and continuing within the Industry that I love. The industry that I am as much a part of as Wendy Williams. In other words, out of the loop. And, I was already making serious preperation to implement my own business strategy. Which, now has been placed on hold. While, God, and the Spirits help me to perfect such a strategy, expanding upon ideas that, bennefit not only Griffin Cleft, and my family. But the Greater Good.

Yesterday's training exercises led me to the first time I quit the process. Well, kind of. Each time I give up, I kind of get stuck in the wondering of "How did they do that?" I know that we're perdominantly speaking about God. So, it shouldn't amaze me to witness his works. However, when you see the grand orchestration that involves people, cars, sounds, Sunlight, Moonlight, Urban Wildlife, weather, clouds, smells, mobile alerts, the lyrics spoken of a song, and so much more, that works in perfect unison. All of it, at the exact same time. Designed to the very second that he intends to make his point. I'm still often beyond belief. It has become so hard for me to just accept the fact that God is creating this world with me, all by ourselves.

Yet, I want to think that the U.S. Government is some how behind a lot of it. I often think that my family knows and is keeping the truth Secret from me. My Mom seems to always call, or make an offer, or say the most perfect things, right as I need them. I've ventured into thinking that maybe my electronics have been hacked by the Catholic Church, and they are creating these very unlikely events in my life. There's a reason for why this process feels so hands on.

As we build the coporate structure of Griffin Cleft. We are creating a business model that is efficient at it's most compact operation. As our ideas, products, and successes expands. The same model expands actively, in real time, mirroring the growth of the company. This allows less interuption to the flow of our creativity during growth spurts. As Mrs. Four Octives often reminds us, you have to plan your growth. You can't just plan to fail. But, you must also control the scale and speed of your growth, in a way that can be managed by you. So as we rehearse Griffin Cleft's corporate practices and processes, the bulk of the work is exploring the immediate capabilities of our efforts.

Of course, that with God's personal guidance and assitance, I will one day be capable of a great many things. Dare I veture to say that, there won't be anything that I can not do? But, that's one day. How does my capability allow us to beging today?

My home/office life is something like this. The Ceo of Griffin Cleft is God, and I am it's Managing Producer. Griffin Cleft is broken into Three Branches, similar to our American Government. The Managing Producer run's the entire Production Branch. Which for me is like being the fun CEO. The CEO that only deals with creation. After, the product is completed. It's someone else's job to figure out how to sell it. And, how to pay for it.

You see how right at the gate, I gave myself a plan for growing. Ceo is a position I plan on earning, and will enjoy several positions along the way. This breaks the glass ceiling that most entrepernuers typically place on themselves. Griffin Cleft has an operating Board of Directors that consists of several members of the Christ Family. You know? Me, God. A few of the Disciples. James sometimes. David most of the time. Solomon, Esther, and Mother Mary all serve in Advisory Capacities, when requested.

Our meetings tend to run long, and full of conflict when you get all the Christians in the same room. Of cource I can't see any Racial Ethnicity. Nor, have I asked. There's no need to. In Mark's world the Original Cast of Characters to tell the story of Christ on Earth, act like a bunch of Black folks. I've heard that they live in a place of perfect clarity. And, that they see sollutions for any "problem" that could ever arise. Yet, when we start our meetings. You got spirits rolling in late with their excuses, not prepared with the work that's due. And, placing all the blame on me. I'll often reply to them, something like. "If my dishes need to be washed, why don't one of ya'll get in the kitchen and bust some suds?" I can feel that it as funny to them as it is to me.

There simply is not enough time in a day for me to wear so many different hats all at once. Especially after having spent years out of the game. Years of not giving a damn about what I or my house looked like. Who going check me, boo? Clearly, my supervisors will now. Ubsurd as it may sound to you. God and I have compartamentalized all of us to respective departments, with specific expectations just as though we were salaried employees. We attend weekly status meetings. We conduct separate wardrobe meetings, and camera rehearsals in which I have a hair and make-up team. Along with a Director and Production Assistants. We have editing meetings in the cutting room.

They train me with music lessons in both piano, and vocal music. I am giving business training in addition to World History lessons. Education in Political Science, Modern Philosophy, Spanish, Japanese, and Russian Languages, in many other courses that comprise my Autodiadect curriculum.

Adding to my confusion, some of the lessons are given to me by very familiar people. The other day, Beyonce gave me a lesson in submitting to the science of what has to occur to reach impressive singing techniques.

She said to me, "I have never sang a single note." Meaning, that she relaxes into the power that is God, who then does the work for her.

Giving back to the community is most important to the Christians. So, each week or month we plan our giving. Whether it be a financial donation of some kind, or my time volunteering some of my highly transferrable skills, that he has blessed me with. Sometimes, my Father in Heaven will tell me that my Mom could use some flowers. I then deduct the cost of the flowers from my weekly philanthropy budget.

We have weekly finance meetings. And, meetings for travel plans. I have several electronic calendars that keep various schedules. All of which, are assigned to and managed by the respective Spiritual entity that is designated to it's department. However, when God sends the bodies to connect focused energy of a position already in existence. Griffin Cleft's new Administrative staff should feel the importance and natural workflow of their tasks. I guess one might say that we are anointing the staff of Griffin Cleft as we complete the preparations for their arrival.

Giving up is also child like in God's process. When you say that you've had enough, he bothers you even more. His bothersom, and mysterious ways will often include flirtation. Yes, girl! God be tryna holler at me. Unfortunately, I know it's because I have been successful at ignoring his other attempts. He knows that making me feel like Daddy's Little Prince will always yield some kind of a response.

All my life I wish to have been a Daddy's Boy, rather than Momma's. Each time I buy a car, I have always chosen the car that I could imagine being held up, at a stop light. Where a young boy standing with his father on the corner, points right at my direction, and says. "Dad, his car is so cool!"

In the American Gay Community, we have many other roles besides Tranny, Hooker, Faggot, etc. I am actually the epitome of what you would call "The In Between." Which is my favorite grouping in the community, btw. We're like Wendy's, Norman. In Hot Topics. We're the perfect mesh of masculinity, feminity, culture, class, and typically are the most fun to be around. The only thing that may seem like a problem to some. Is that our well elocuted speech, and fiercly energetic walks usually provide, most of ya'll, enough evidence to arrive at a label for our sexual preferences, before we say "Hello!". It was the perfect way to be in Los Angeles.

Now, in New York, unfortunately. If I forget to act my race, while desperately trying to remember which leg I did my swagger limp on. And, even harder todetermine which side my cap should be tilted towards. Not to mention, and it's a big one for me. Trying to contain your composure when running into a friend you hadn't seen in a while. You should never show as much excitement to see a person as I do. Okay, lets see. Be unresponsive to all compliments, like the jaded and stressed streets of New York are overflowing with complimentaries. Yes, and last thing, but the best thing you can learn about New York's Gay "Community." Don't you ever, ever! Say that, you are Gay.

Make up any reason for why you are a Forty year old man, living alone in a One bedroom apartment. Sparsley furnished, lacking of any items in view that gives way to your true personality. Come on man, really? You got a Patrick Ewing poster up on your wall in 2015? Na, Un! Bitch! You need to get some new Knick souveniers if you wanna pull off ultimate Knick Fan of the Season. At least have the current schedule and ordering form displayed prominately. Them shits is free, every day! But, I forget. Here in the city where Blacks and Gays don't buy into their own facades. It shouldn't be that uncommon for them to put forth lackluster effort, in staging their persuasions. As if we've all aggreed to a city wide ban on Second dates.

The quality of life that our Black Gays are experiencing in New York is so sad, and lacks expanding qualities. For instance, You move into a new home. Have you asked yourself, How would you like to feel in it? I'm certain that you have asked yourself what you'd like to see present in it.

Even then, most of them would say, "a couch. I could put a lamp right there." And so on. But, the minute you can add to your description, "I see a Blue, Sofa of Loriot, tuffted in pearl, with added on Goose Down Sofa Cushion Toppers to raise the seating height, which sits across from a set of Two, Pearl, Louis The Fourteenth Parlour Chairs." Just by adding that bit of detail, you have already increased the quality of your life. Simply by saying that for your living room design, only specific pieces will be suitable.

And, cancels out the "anthing will do for now," energy that the bulk of the Bueens in New York are generating. Sometimes it can feel a bit like we've all just arrived here, or. We may be leaving soon.

It's funny that until now, I haven't been able to put my finger on why New York feels like worlds removed from the New York I once knew as a child. Time spent discussing the possibilities of Mark's World, with God. Has allowed me to pin point some of the reasons I could not agree to remain living here beyond my transitional period.

I knew that things such as politics, strength in my local Gay Community, friends just as excited for creating, as much as I am, and a few other must haves, are important for my solitude. I use to feel quite important, when I made decisions from the boards that I sat on in Los Angeles. For absolutely Free Ninty Nine, you as a gay men can go join the efforts of your local, Gay non profit charity and gain all kinds of perks, connections, but, best of all. Some power. God once told me that there are Two types of Powers, Money and People. With one, you are respected and with them both, you are feared. But, lack them both, and you are oppressed.

Can't believe how it all is coming back to me, and rather quickly now. I spent half a decade waiting patiently. And now, not knowing how long I'll have my Christian helpers. I suppose I could just call them Angels, for argument's sake. However, Angelic to me means with the lack of purpose outside of God's. As we explore this spiritual relationship of give and take. God reminds me that this life has to be my choice. He certainly wants me working on the asks that he has given unto me. Yet, carving out time to do my own thing, is just as important to him.

He also, reminds me of certain dreams that I'd hope to make come true some day. Dreams that once made me really, really, passionate. Still, I managed to forget them as new dreams developed. He continues now to allow the new and the old dreams expand into the overtures to strategy, processes, and identities that surpass what I ever drempt obtainable.

By the way, I almost forgot to tell you. I identified several true believers who I were certain of. I called them up and asked them to quiz me on the Bible. I begged them to give me something hard. But, nope. I couldn't be stumped! The idae seems possible to skim through the surface of Two

Thousand years of history, while simultaneously combining your own understanding of the gaps where no study is required. Especially with the resources, allowng quick access to the worlds acceptance of what is written within the Holy Pages. Yet, wrapping my head around the idea that I nor anyone else could have ever read the Bible in Five hours explodes my mind! Yet, the word on the street is I am number Five.

I got my first job at Thirteen deliveing the Staten Island Advanced. I worked there, and other jobs continuously through out Junior High, and High Schools. Once my Ballet training began to look like money and time worth while spent, I traded in jobs like the New York Urban League, where Jennie Smith Campbell took a risk on a young, and bright Fourteen year old to chair her Youth Initiative Council. To jobs within Concert dance.

Jobs such as, The Rod Rodgers Dance Company. Or, moonlighting as a special events dancer at the Winter Garden Resteraunt in Brooklyn. In fact, there was a time when I knew more Russian than Japanese or Spanish. All of our rehearsals were run in Russian only. WIth the exception of Two other guys and myself, the entire production and restaurant staff were from Russia. Our choreographer did not, in fact, speak English at all.

Yup! Right here in America there exists so many ghettos , housing foreign travelers so comfortably, that our visitors can stay for decades without bothering to learn English. As well as work full time jobs, build a family and sizable fortune. All while flying under the radar to America's knowledge of their existence.

Over the next week so much transpires that I am not sure how to itimize it chronilogically on my timeline designated to The Purple Period. You're familar with my formal education. If not, let me catch you up to speed. I ain't got none. So the scientific concepts that are introduced to me are all foreign, and for the most part I have conducted no individual study on the subjects.

However, as the voices that once chanted, "You can't get it wrong, you can't get it right." Now chant, "Remember." So much so that, now I'm often drugging up the past, into my new Process. Especially, boring lessons that I didn't want to learn anyway. That were taught so badly in grade school, and Sunday School. And, even more difficult, to remember details of a plan and purpose that you were instrumental in creating before you even came to Earth. Again.

But, what God did in my time with AC, was comprise all of the review that would get the squeaky wheels of surpressed memories lubed up again. So lubed up, that memories are provoking furocious debates between AC and I. Heated would be an understatement. Not, arguments, however. We wouldn't hold anything back. Especially me. When he contributes one wrong piece of information into our exploration that validates our knowing. I would chew his head off, with name calling and shade that not even, shady Phaedra Parks would sanction.

Interestingly enough. Nothing could offend AC. Except the one time our debate got so "heated" that I threw a chair, his direction. I really wasn't aiming to hit him. It's just, that we had only gotten the chair the day before, and I didn't know how heavy it was. So, yea. Other than that hurting his feelings, more then the hurt he would provoke himself with, wasn't possible. Packaged with the lowest self esteem an underserved Southern Public School Education could create in a Man.

Vague is not the word, when you start asking him specific details about his very recent life. Eloquent stories of his childhood would be intimately detailed. However, He has lived in New York for Three Years, but doesn't know the area's township title to where he resides. North West Harlem to be specific. Since meeting him I have had to give him his rent money twice. He is renting weekly, so don't think that my broke, cheap ass paid his New York, Two Months worth of rent. No manm!

He also has given me several different stories about the type of place he lives in. As well as who owns it, or even a correct address for that matter. Oddest of all, is that I couldn't understand why someone as talented as he is. And, calaims to have made the voyage to New York, all the way from the family whom loves him. In the name of dreams. The pursuit of carving his section of fame within the Fashion Industry. Would then, be stuck in a dead in job waiting tables for Three years. And, had never conteplated the idea of quitting before he met me. A job that wasn't even full time, no room for growth, low paying, and a great distraction from the ideas and strategies that accompanied him on his Journey from Georgia.

Not sure how he has been managing his money, he is now asking me for his rent money again for a Third time. As I refuse him, he becomes some what different. Not from a place of anger. But, from a place of shattered stabiity. As if he had considered me to be his financial provider in all ways, unrelating to his salaries and advances.

I gave him very favorable terms when I hired him. Especially considering his lack of a resume, experience, and capability. It just seemed as though he truly were my Soul Mate. However, I had to draw the line at constantly paying for his personal overhead. Literally. If his worry and un appreciative attitude and energy wasn't enough. He would never want to go home. Another challenge working from home presents when having staff in your home office.

Since he really lacks in the administration of Griffin Cleft, I altered his position to become more of a personal assistant. Which he gleams joy from. Me on the other hand with a start up that I've invested every penny into, can't justify the line expense for a personal assistant. Who am I, Beyonce?

It wasn't my idea. The orders came direct from my CEO. I enjoy this planning of gowth, processed into an expansive evolution that appears logical. Non-threatening and includes hap hazard reminders of, capabilities I was once effortlessly advantages to. For, example. After loosing all of my valuable assests, I worked so hard to acquire in California when moving back East. It was a most welcomed and sweet environment when I moved to West 137th street. The obligation of only being responsible for a single dorm-like room, lacking a kitchen, or bathroom even. The careful reintroduction into a culture of maintaining one's own apartment, household and business dealings. Then to move up to a one bedroom apartment at the East 141st Street place. While holding on to the processes, and disciplines Igained, at West 137th Street.

I'd say, that thigs are going pretty well for Griffin Cleft and I. The ease of expanded ideas that didn't seem to differ all that much from what I did before adding to the ideas. However, I could have never anticipated what would come. I have lived my entire life with the expectation that I would be greater than most.

It is not odd at all to me that I have been chosen by God to do his will on Earth. I even look forward to all the crazy channeling, and training exercises that often leave me feeling like it's all bulshit at the end of the day. What could seem like children's games, are processes enriched in lessons that will equate to valuable tools, in future strategies.

As AC retreats to a corner, to try drumming up some clients to style. Praying that, that will be the answer to his rent being paid this week. I continue getting to know all the wonderful technology that God has blessed me with in my new home/office. Probably so common to most of you. The highlight of my day, today. Is downloading the Pandora App on my new Smart TV. I've not owned my Pandora account for all that long of a time. But, getting back to the selections of music I listened to over and ovaer again as a kid, is a special treat for me.

Growing up, I was the odd kid who knew nothing of Pop Culture, and only listened to Classical Music. I would spend most of my weekly allowance, to purchase Taped recordings of Classical Masterpieces. When I look backwards at my life, I am impressed at what sophisticated taste I had in music at such an early age. I mean I wasn't listening to bullshit Mozart for Babies. Or the popular romantic ballets that would become some of my favorites in the future. But, from the age of Five years old I was listening to Brahms, and Shostokovitch. Mendelson and Bizet. Lakme, and Orff.

In fact, if Pandora hadn't reminded me of some of the brilliant music that I loved as a child, I would have told you that I didn't like compositions from the Classical period. Now, and for the past Five years. My love has mostly been for the Baroque. Especially now after discovering the Italian Baroque, outside of Vivaldi. Vivaldi's heavily German influence does not reflect the romance present in a composer such as Arcangelo Corelli. Which at times if you closed your eyes, you could almost hear R & B, base lines that would not have been appreciated at the time concieved, like I can today.

Reconnecting with my childhood favorite works on Pandora has an added bonus. Modern times now allow us not to just listen to the radio, but to watch the radio. Being able to read the exact details regarding a recording as well as biographical information along with it not only gives a new respect to the knowing of the works. But, also learning similarities of the composers to one another. And, learning information pertaining to the exact time periods of when the works were created. And, the purposes for why they were created, is exhillarating to me. Amazing, isn't it? At a time where there were no recording devices other than ink to paper. Humans were able to create brilliant works of musical art, that would be as relevant now, if not more, as when the works originated.

My personal growth, has individual studies that I conduct every year with God, with in to increase my academic knowledge. It's kind of our thing, since learning the historical origin of anything has always fascinated me. So this year, I have finally arrived at the study of numbers. At this point, we are studying the number Five. Understanding the purpose of the creation of numbers, draws my attention to numbers in an enlightend manor. Also, not believing in much coencidense, I begin to study the numbers of the classical works. Classical music uses an antiquated cataloging system, that works perfectly in our modern day.

Each section of a work is given it's section title such as, "Movement 2, Adagio." Then those sections are catagorized, in its entirety with a number that coencides with the order in which the piece was created. Such as Symphony Number One, bwv347. The acronym changes depending on the reigon of the world that the composer is from. Bwv is the most common, since used by J. S. Bach, who has the largest music catalogue of any composer.

So now, that I can read the bwv number adjacent to the title of the work. I recognize that there is meaning in the titles of the works. Almost, like a secret coding that only people who have been chosen by God recognize. True believers that actually took the time to learn not just the origin of a word or number. But, it's purpose relating to God's needs. As you can imagine. To God, only his defining purpose matters to him. Everything else is just bullshit in his eyes. But, he tells me that, it is how we should also view the defining purposes of our own lives.

Keeping in mind that with anything done to the extreme, abuse is enevidable. However, if the rest of the world sees your ladder as a desk chair. Fuck the rest of the world. If in your life the only purpose you have for that desk chair, is to be your ladder when reaching behind the refrigerator. It's a "ladder!" Move on. It's not narcasitic or crazy at all. For there once was a time when a desk chair was just a chair.

The shit that went down yesterday hurt especially because I felt as though My nephew's death and the change that it had prompted in my life. Was used to manipulate emotion in action in me that felt overwhelmingly invasive. I document these events with better clarity, now. But, when the Purple Period began. A grasp of who was doing what, to yeild what change, that would effect who, was far out of my reach. Each day God adds more to the story of why I was chosen or what I was chosen for. One day he told me that I was helping him create a new world. And, that the New World would be one of my own.

Pondering that idea, I fearfully drift off into sleep. When I am consicous, I can do a pretty good job of staying present while ignoring the voices, process, and worst of all, God. But, asleep? With my body being in a total involuntary state of being. My innerbeing refuses to stop hanging out with all of them in my dreams. Normally, I don't know who to expect a visit from in my dreams. Like I mentioned, often old women trying to reach out to their "Momma's Boy's," from the other side. But, these days mostly just him. Also, The Lord and AC. AC and I do a lot of communicating with one another through our dreams.

A question that is constant in my mind, is how long God's process will last. Was it something that I would have to eventually let go of,the next time he choses the class of bright autodiadects to teach his Process to? So concepts like communicating externally in dream state made me wonder if I were only experiencing such ability because God was present. Would that ability dissappear when he left me? He's made it quite clear that I am becoming a man.

Clear that he would be traveling with me on a Journey that would take me back to where I came from. The idea of God dropping me off like it's hot, and returning to a world so far from me that faith would be our only resource for validation, was a bitter sweet, but a real conern of mine. I have to say that I fell in love with the guy. Undescribable to try explaining how the love of God feels. It's a love that you understand right away will not ever be a love any human can accomplish on Earth. Although he says we can do whatever he can do.

At any rate, when he would leave me for the day, I would miss him. When I would awake to a Substitue Teacher, in lew of God. I would also become Sad, and angry.

I wanted every waking moment of mine to be filled with the magical workings, that is God's process. Sometimes days would go by with no word at all. Then, I would have an unpure thought that causese inner emberassment. And, his voice would let me know that he had just experienced my unpure thought. Reminding me that even without feeling the energy of his presence, he is.

"You need a wife!" This is what is interupting my sleep this evening. What's worst, is that I probably wont even remember having the conversation with God tomorrow. But, it goes on a little something like this. "Being a part of validating American presidencies is a great honor. And, this time having your president be Barack Obama. We can finally move forward from the extended interuption that earth has caused in our universal design.

Human civilization began about Two Thousand Years ago. And, it has taken you this long to migrate from the shear nothingness in which you arrived at, in the African region of Earth. Through your migration, intergration, and revollutionary feats, A man with roots held strongly in the African Reigon, and is of Mixed race of European, and Africa Ancestry. He now rules the Free World."

I got it instantly. It helps that in a dream the resources for visiual aids are endlessly abundant. So, you can imagine the underscore of African Pride Drumming I hear. If, we were setting this story as a movie. The sweeping transitions of still and motion shots of beautiful serengeti landscapes would remind you of. Wait. Let's say it on three. One. Two. Three. Mufasa! Totally. I felt as though I were baby simba, being shown his Kingdom for the first time.

OWhen AC and I began in Genesis. The description of Earth didn't seem

The Purple Book

all that different as some of our planets, floating out there now in space. Particularly, my favourite. The Planet Mars. It states that the Earth was full of void and waste. I'd say depictions of characteristics similar to Mars and others currently.

To understand his intent in describing such a place, I looked up the definition of Void. We can pretty much agree upon, what is meant by waste. **Void:** not valid or legally binding. The term valid implies that there is some measurement of set qualitative results. If true. Who set those measurements? Who facilitates examination, and plays the deciding role in the conclusion of validity?

We could simply say that God has the Audacity to make his own defining judgements. Quite often when I am no where in the vicinity of finding the correct explanation of why something is valid. He will often say to me in a loving, fatherly tone, "Because I said so."

As the dream continues. God explains to me that with Obama's success, we could all die now. We've done it all. Everything that could or would, and many. Many, times. God is so tired of everyone wanting to live for ever, and participaing in the culture of stuff. The number of people ready to retire their spirits to live forever with God in Heaven dwindles daily. It was even part of my reason for not wanting to become a True Christian. I understood that when True Christians die, they live eternally in Heaven with the Father. That was really of no interest to me. And, sounded quite boring.

The attraction to the newest of worlds where we can get the new 3016 Mercedes Hovercraft that flies you around Mars in the comfort of a luxury sedan. The excitment of the needs and requirements for a world wide migration of Earthlings to Mars. I was in love with the idea of playing a historic role in the migration. Imagine. Me. U. S. Secretary of Defence. Or the director of NASA in 3012 when we begin the safe migration and transition to coexist in the new world of Marshmans. The "H" is silent.

But, no. God assures me that there is not a need for any massive migration to Mars. Nor, do I need to continue coming back to Earth, life after, life. Simply to assist with mantanence of something that is done and over with already. Imagine that you get your hair done at 10am. Then at noon, someone makes a house visit to rebump your wrapped bob. Then at 12:15 another comes to wash the curls out. But, doesn't stay long enough to complete the dry and professional hair styling, that they washed completely out.

You're now left with the choice of improvising with your amateur capabilities. Or, go back to your stylist who may have the time to fit you in the day, that they have already begun moving forward in. Only to take backward steps with your, styling emergency, after they just styled it hours ago. Then you ask, can I get a discount, since it's twice in one day.

How is that fair? The dramatic inconvenience of connecting you back to the reset where you are exiting the stylist's salon with beautiful, healthy, showstopping hair. For the second time today, I might add. And, now after obliging all of your requests. And, adding to the work load that is now put your stylist behind schedule. You want a discount? When your request for a discount is refused, you say that you feel that you were treated unfairly and must go to another salon in the future. So here we are. Stuck with the redundant task of creating more Hairstylists for the purpose of you not wanting to keep the perfectly good stylist we just provided you with Twenty years ago.

Worst of all. When the starnger shows up to wash your espensive hair-do out at 6pm. You forget that this has already happened. And, that it probably is best to just say no to the stranger who only wants to create what is already in existence .

I hope that concept makes since to you. Because, I got it right away. The way that God describes the starting point for Humans inhabiting the Earth, really suggests to me that in creating it. He was only rebuilding what he had already destroyed. Again. I'm no genius. I just got clarity when he explained that there was no reason to really want to go to Mars.

I would arrive in the abundance of darkness, waste and void. Only for Hundreds of Thousands of years to pass as we create a mirror of what Earth once was. And, finally nature takes it's course with the death of a new planet. Followed by the migration to yet, another planet to then begin creating the mirror of Mars. Of course, starting yet another time. Here we stand in the midst of darkness, waste, and void.

We could probably piece together stories from worlds beyond the grasp of our Humanly, 2016 brains can comprehend. But, they tell us that we only access Ten percent from the source of who we fully are, while on Earth. Which would confirm that Ninty percent of who we are, exists some where else. Which begs the question for me. How much of us exists with God?

If any of it does, it's hard for me to find relativity in God's need for more conpanionship in Heaven. With the Spiritual Equity that my Ancestors alone have paid for. I would think that we would have a town named Griffin Quarters, somewhere in Upper East Heaven. If not, where the hell are we all going?

Forgive my becoming off topic. Back to the dream.

Knowing that God is real. You must understand the constant threats from his enemies. He warns me early on, that he hopes to gain my trust in his process. But, cautions me not to ever trust him completely. The idea that we can do all on Earth that he can do in Heaven continues on the other sid, unfortunately. Meaning that others can pretend to be him. Which is why much of what he tells me, is untrue. However, it's what he needs the mimickers to overhear, in the cases where they ease drop.

And, like a Mob boss. He's dips and dives, while popping in and out, unexpectedly in the most random of times and places. This time, about an hour into my good sleep. After a very long couple of days.

God is meeting secretly with me and others chosen, in my dream that night. I don't recognize any of the faces besides AC who is also there in the dream. Some of us are divided into groups of Three or more. But, our group only consisted of AC and myself. I suppose we had Three as well if you count God as a participant in my group. As I write this now. I feel as though, that if I were reading this book, I would think that Mark adapted, with a great handle on what was being thrown his way. But, at the time I thought the only common ground that would position AC and I on the same team is that we both wanted to go to Mars.

I assume that we're all Christian since God reiderates in the dream that value of all the tasks we complete, is most valid because we are the True believers. "In order for you to even become close to succeding in my process, you must have more than a mustard seed worth of faith. Without the total belief in your purpose on Earth, your new world beyond Earth lacks the depth that scales great value. Value and purpose is now your only reason for action."

Naturally, hearing the word "Value" sends me to think in terms of money. His next statement reassures me that I am correct and thinking that this has to do with money. "As a result of you accepting and being accepted into my grand design, which bennefits the greater good. I am making you a member of, The One percent."

After only accepting Jesus as my Lord and Savior a week ago. I have resitance to continuing the journey that would take me deeper into religion. However, trying to build a company this second time around. I need all the help I can get. He promises that he would get me that help and the finances to implement the start of Griffin Cleft. And, he promises that even in my failure I would have a successful company. That's quite a bargain that makes the risks that accompanying his requests, seem almost tollerable. I pray that it is.

He continues to explain in the dream that AC and my realtionship is now growing with a concept that is unfamiliar to the both of us. We'd grown to love one another. There is a bond happening, that would appear as though we were a romantic couple. However, we lacked all feelings of romantic attraction. But, something more special than a bromance.

He tells me, "Most everything that you do from this point on, you will do at least twice. This tells me that you really want to achieve the results your actions could gain you. Specificly for Mark, you must lower your resistance to getting married. AC, will not make the cut for a suitable husband. We can use him, however, for practice. Eventually, preparing you to be a suitable husband for whom I've prepared for you."

Until now. I have always taking comfort in being a Gay man, that enjoys the release of liability that accompany's having children and marriage. Politically, I'd hope that Mariage was not redefined in anyway. Even to include civil equalities that should be of benefit to me.

Firstly, I have true appreciation for origin. Participation in creation, I feel, is not redefining the things that existed long before we got here. When Marriage was created for the purposes involed in increasing the worth of an estate, no one stopped to factor in the day where men might want to marry other men, and create estates of their own.

So, too bad. Living in the Free World, I felt that the opportunity to create something better than marriage was misse. Now, that we are conforming to the practice that originally excluded us. The opportunity for Straight Men and Women to say, "I wish that we could become signified like Gay people now have. It's so much more logical than marriage."

The way that you best enjoy rituals, should be buy standing in the truth of the ritual. The ritual that is marriage, is not how Gays should unite. Since most of us bueens are lacking in strong family ties. And will be starting, in most cases, the first generations of ourfamilies. We deserve a ritual that has the essence of fresh and new. Not, one designated for uniting Two historically, and endowed families with one another.

In this case we not only, truly take on the roles of creators. But, we also introduce a binding experience that trumps the exclusive marriage club. But, since we didn't. Now my primary task is getting married. Me. The guy who doesn't understand how people choose just one. Married?

There you have it. Some other Big Dreamer beat me to the punch. And, now that he has altered history, with his federally mandated admendment to Marriage. Gay Christians can no longer. Well, not in good conscience. Continue a culture of shifting partners like chess pieces, resting on the illegality of the alternative.

In my formative Christian years, you know. The first go round. I always dreded the Sunday mornings, where I would show up just in time to get a front row seat to the Annual "Homosexuality Is The Worst Sin Sermon." I wish that we could get a newsletter or something alerting us to when the Gay Bashing Sermon or GBS wil be scheduled for. That way, the long time members who have heard it all before. Like myself, could opt out. I typically leave when I find out.   At least an alert would allow me to sleep in and skip it a year or Two. Hopefully now, that with the support of the American Government, President Obama, and Pope Francis. Spiritual leaders will begin swapping the GBS, for the Marriage Sermons they once refused to perform.

Throughout the dream we speak more about what my obligations would intell. Certain promises that I would have to keep, to stay in God's Grace. And, The Lord's Mercy. His request from me, didn't seem like much of an effort on my part. Asks like protect the children and never molest them are ideas in my core value system.

So my growth is not some monumental, life changing event. It is in fact, a series of small, time sensative adjustments that include defining commitments. For Christians, the consitency of honoring those committments allows God to trust that work he will now conduct, match your future within his Grand Design. And, wont be wasted. Nor will the breaking of such committments allow destructive disruption to a massive construct that effects all that is.

The next day when AC arrives, I mention to him a bit out the dream I had last night. He tells me that he knows and that he was there. My mind now wonders off to the imagery of the movie Inception.

For most of the year, against God's better judgement. I was in the costant battle of trying to figure out how the magical elements of God's Process were done. For example. AC and I were coming home from dinner in Harlem one night. At the 116th Street Station, on the 2 Train. The time lapsing until the next train would arrive said clear as day. Ten minutes, as we entered the turnstile.

Just as we enter the platform, which takes the whole of Three seconds, we look up to see, that the clock has counted down Eight minutes and now lists Two minutes as the time before the next arrival. Accepting that God has the power to do all just wasn't enough.

Eyes would move when I looked at posters for coming attractions. I saw Three Squirells run up the stairs to a Town Home Porch, and then morph into pigeons that flew off into the skyline. Many other events such as these kept me in the wanting to know how it all was happening.

Of course we've all heard the water into wine stories. But, who truly believed? Especially not expecting the day to come when I would witness similar, unexplainable miracles of my own. I sought out for logical reasoning that would validate the Science behind my experience constantly.

I, sometimes fell short of finding reasoning, due to so much newness being undiscovered by the Ninty Nine Percent. Even the great American Scientists have given up on validating religion, or the power of it. Now left with the only optio, that is trusting God, comes with his confusing instruction of, not to trust God. Instruction that he himself demanded of me.

Stuck in the amazement of it all. I would often here God say, "Stop trying to figure it out." There were parts of me that thought that I was truly going crazy. Mental Illness does in fact run in my family on my Mother's side.

Once, I stoped to look at myself in my Mom's rearview mirror and thought. "If a stranger on the street passed by just now. They would have thought that I would have Skitzofrenia." My uncle does. I've watched him have heated arguments, to joyous reunions with the people he spoke to. Yet, no one can see who he is talking to. And, just then as I spoke aloud to God. And, laughing hysterically. I noticed that if perceived by a stranger, my actions would warrant a similar assumption as to my Uncle's condition.

AC's conviction in Christ was really admirable. In his life, Christianity was the only solid. So many in my family who witnessed my Christ like behavior, tried to convince me to give my confession of JC as my Lord and Savior, to no avail. I didn't harbor any resentment towards Christ. But, I was not in the habit of continuing the practice of religon that had failed me in my previous attempts at Christianity. But, discovering conviction so strong and pursuasive as AC's faith in Chris, made it a no brainer to accept his logic. Then, let Jesus go ahead and take his rightful place in my life. Trusting that this time around would be different as he would guide me to better understanding of his Life's work.

The GPS/computer chip was one of many abilities that appeared for the first time in my life. AC was convinced that God was giving me super powers from this process. I continue to be apauled at AC being present in the moments that I learn how to tap into my "Super Powers." But, he would walk away, gaining no more knowledge then he arrived with. But, he was determined to be as powerful as I was becoming.

Each day, I would tutor him in the process that would create the New World. But, like a todler, he would forget the lessons that he learned the day before. There were mornings where I would conduct a tutoring session and by noon, he would have already forgotten the evidence that would support science we were exploring.

Most often, AC would be given clues that would totally go over his head. Then I would piece those clues together, leading me to revelation that would knock my socks off.

Although my trust in God is still a bit shaky, I am in love with his process. When I tell you that he thought of everything. He thought everything. I could hear God when I was awake. But, AC could only talk to God in his dreams. Which was an amazing "Super Power" to witness.

Whenever I would get close to a discovery, but be stumped by a missing puzzle piece. AC, would literally lay down, fall right away to sleep, and wake up a couple of minutes later with the missing link that would advance us to the next level in the curriculum.

Sometimes I would be like, "God I know you are standing right there watching us. You couldn't have just told me that in my ear?" How is it fair that AC gets to control the destiny of my God Education? He doesn't even understand the information that he is passing along to me.

AC would get so jealous of how quickly I would grasp the knowledge that God was imparting on us. He would study so hard for the quiz's that I were being developed by Griffin Cleft. Yet, he failed every single one of them. Especially, the new math. God gave us a goto answer for the times where we just couldn't remember how we arrived at an answer. "Because, God says so!"

AC would spend so much time on a single question, struggling to understand what was being asked. When he could have answerd every single question with "Because, God says so." He never gave us a limit on how many times we could rely on the goto correct answer.

AC has actually had more education than I have. But, I swear that God presses a reset button that wipes his memory clean of anything that isn't rudementary common knowledge. It was cute for a while. But, I am growing so tired of wasting time with tutoring him on content that takes him forever to comprehend. And, tomorrow he won't remember any of it. Yet, I still had to wait for God to deliver the clues to him, in dreams. Before I could move forward, exploring the validity in our findings. It feels a bit like torture since God has chosen him to be a liason to the missing links in my guidance system.

Aside from our studies. AC and I were starting to gel into an old married couple. He would handle all of the "wifely" duties, such as Grocery shopping and cooking. All of sudden, he no longer let me come into what use to be my kitchen. Only when it was time for me to empty trash at night. Yes, I played the husband. I treated him like a prince. Taking care of his every need.

I would even find myself buying impulse items that just reminded me of him. Then, would look forward to seeing his reaction after I gave him his surprise gifts. But, with all the elements present to make up a wonderful marriage. We still hadn't developed any romantic feelings for each other. Or at least I hadn't.

I'm not sure the same could be said for AC, anymore. I hate the idea of dating anyone who towers over me, making me appear to be shorter than I really am. So typically, that's my deal breaker. For him, he didn't find me to be butch enough. That notion was quickly being tossed to the curb.

You see, I knew that we would be filming Mad Science this summer. So, I was no longer laying around the house, eating unhealthy snacks all day in second hand, drift store threads. I was conducting daily skin treatments, debuting a fresh haircut weekly, and hitting the gym hard. In other words doing all the things we performers have to adhere to when becoming camera ready.

As a stylist. You get the opportunity to create the best estetic on a person, imagined by your own preferences. So as he dressed me in styles he was attracted to. He began to change his mind about being attracted to me. He began making advancements towards me, and constantly reminding me that he was ready to take our relationship to the next level. Oh damn!

I knew that we were heading in a risky direction by this point. Wish I could tell you differently. But, he had not made similar efforts to improve his looks in the way that I had. I couldn't share in his newly found attraction in me. The deal identifying our individual roles, that had been working eficiently seemless. Now took a turn for the worst, the first time I rejected him.

I don't know what he expected, when I couldn't be any clearer about us not being ideal romantic partners. He had the deal of a life time, I thought. A job, place to live, companionship, a man that treated him like a twink boy toy! A man that was the same age as him, not twice his age the way your typical Sugardaddies would be. The freedom to smush with anyone he wanted to. He had it made! But, some people really believe that they can have it all. Becoming destructive to everything that is succeding in their lives, simply because of that single desire that is not being met,. Or, could use some improving.

As you could imagine, destruction became the path taken by AC. From this point until the end of our relationship he became posessive, needy, and mean.

The final week of my time with AC was the wierdest of all. The night before, AC asked me if I could try his power of asking God for answers when sleeping. He was struggling with some issues he needed solving. So I did. Eventhough my dream that night was full of coversation, battles, travel, and anything else that would leave a person feeling fatigued once they awake from a full night, of what should have been restful sleep. I woke only remebering one word that God wanted to tell AC. Just like my clues relayed through AC. The word flew right over my head.

After I woke, I called AC and asked him to come over. As I ended the call, my phone rang. The caller ID showed me that it was my front door. I let them in without asking who it was, because sometimes the children in the building would ask me to let them in. They're parents won't connect their buzzers to their phones. When I got the knock on my apartment door, I wondered who it could be.

"How did you get here so fast?" I asked. "I never went home." AC answered. He continues, "Last night I didn't want to go home, so I slept outside your door until your Security kicked me out. After, that I slept outside on your door steps."

Can you imagine how hard life would have to get for your best option to be sleeping outside, on someone else's door steps? But, AC's living situation had gotten so bad that he did not want to go home. "Come on in." I tell him as I looked into the face, that lacked life. Distraught was an understatement that described all that was AC.

With the sounds of a single word life rentered his body. "God said to tell you Purple." After he heard me tell him that, he jumped up and down thanking God for the sollution to his prayers. To this day I don't understand why that was the one word that would comfort all of his worries. "You're going to Mars." He blurted out as he calmed down. "AC, it's too early." I mean! Can I at least get some coffee in me before we start the debate of where we are going. "No, hear me out. I promise no arguing today."

Keep in mind that the dream I had last night has me exausted this morning. But, there is no stopping AC when he has a point to get across. He reiterates, "You are going to Mars. I was already your co-pilot the last time we built Mars. When you arrive on Mars, you will have to do everything that we already have done there, all over again. I wondered if when you went back to Mars, if I would be co-piloting you again. I'm not intersted in repeating history. I'd rather pilot my own mission to build on a new planet. Through you, God has just told me that I will not be going to Mars, again, with you."

6 TAKES ONE TO KNOW ONE

Today I had planned to have lunch at my Mom's. I had to cancel on her because I can't have an employee of mine sleeping on the steps of my building. We need to fix his living situation, ASAP! I've told him that he can stay one week with me while he looks for a place. I would rather pay for him to stay another week at a similar rooming house. But, since they're all full, and he is adamant about needing to leave, and has admitted to not using the rent money that I gave him to actually pay his rent. Not to mention, moving a guy in who was in love with me, probably was a bad idea if I weren't going to return the romance.

Missing lunch today with my mom really sucks too. I haven't checked on her in a while, and. I have a burning desire to get some information from my childhood that can only be dicussed in person. That bit AC told me about Mars really has me wanting to know the source of my fascination with the planet. There really shouldn't have been a reason to not go, since I could have gone and been back now. But, instead I have been waiting all day for AC to text me his address. With all his stories not adding up, I wanted to check for myself. So the plan was for me to pick him up to help him move his things in.

I'm a little concerned that I haven't heard anything from him. He was back to sounding distraught when I spoke to him this morning. I feel bad that he is having such a hard time in life. Especially since I will have to fire him as well, this week. I know, terrible right? He has been doing a less than average job. I'm not balling enough yet to have someone on my payroll just to talk to God for me. The Almighty Father is going have to come up with a better way to feed me clues and information.

Our plans were discussed and agreed upon. However, AC has shown up at my door, instead of texting me his address. He said he had just left a meeting with his agent to discuss a client this week. "Where's your stuff?" I asked since he had only a couple of bags with him. "Its back at my place," he answers. He contiues offering me reasons and excuses as to why we couldn't go get the rest of things. Yet, he still was ready to start living with me at the present. Odd I know, but I have come to expect nothing less of AC.

As we fell off to sleep last night. It seemed as though we were on good terms, still. I hadn't let on that his job was in jeopardy, yet. But, that doesn't mean that God didn't let him know what was up as we slept. Remember how I could remember vivid details about my dreams? Well I can't seem to remember much anymore. I don't know if I'm just not having dreams, or if AC is absorbing some of my special gifts. Whatever the case, I can tell he knows something. Today, his energy has gone from appreciation to passive aggressive.

"Didn't you say that you were going to take me to meet your mom? It would be nice to see Staten Island, since I may have to go back home to Georgia this coming Friday." WTF? Of all days for him to remember that offering uttered so long ago. I'm not going to take this lying, using, stalker to my Mom's house. "You're going home for a visit?" I reply. I already knew where he was casting his line. "No. I probably should just go live with my Mom. I'm not going to find another place to live in a week. And, I have no one else to stay with." Of, course. Didn't you guess it. All day long, he's thinking of ways to get me to change my mind. And, of course I am combating him with calculated avoidance. He doesn't even realize how well God used him to train me in the art of dispute.

When I look at this young man. Whom bares talents as vast as the eye can see. Can turn the heads of guys in my hood, that I didn't even know were living Down Low lives. Knows the ins and outs of Christianity, through and through. Has been positioned in my life truly for equal instrumentation of our individual success and purposes. Has all the makings of what I would expect my soul mate to be. Our source energy is so connected that when he moves, so do I. And, I kid you not! If I am outside as he approaches my building, I always know that he is coming.

It gets so windy. Forcefull winds, that create a magnatism so strong that it literally feels like we're being pushed towards one another. So, it hurts my heart to notice the blessing that God has bestowed upon us. And, he can not. While I relish and the spirits of success and happiness. He is crippled with the dispair of worry and disatisfaction. The end is enevidable. The Lord says that of all the desires that construct our state of mind. That, in the case of Lust. You're not suppose to stand and pray at it. Nor, bargain with it. Or test your will power, or progression. You are to flee. Run! Scoot! Scadadal!

Scram! However you can, if you can get up out of there. Get up, and get out of there!

Because, even he recognizes the destruction that comes from it, when most men. Accepts her challenge. I know some of the women are thinking, "Why lust got to be a her?" Because! It's Biblical. Sanctamonious triumph always get's the masculine posession. While, everythig ratchet, or having to do with evil is always, givien the feminine. With the exception of our Blessed Mother Mary. That "Virgin," who miraculously defies logical, evollutionary, biology to bring the son of God, to Earth. Yet, we celebrate the Life and Death of her son, But. What do we do for her? At any rate.

I just wish he could focus more so on the positives instead of giving up so easily. It's not a luxury that I am privy to when we're debating some fools argument. He once argued the view that non-profit organizations were not corporations because they couldn't make any profits. What a lot of people may not know is that non-profits are very profitable. It's just that plan and raise the financial needs before the work can begin. While of course the for profits do all the work first, praying that the work will be met with extreme profits to report. I'm telling you I had to conceed, just to not wast any more time that night. He probably still thinks he won that one. Feel sorry for him if he ever runs into Walter Liedtke.

You know, as I'm thinking back. AC offers the exact evidence for a major issue I have within the Christian Faith. We practice redirecting our victorious praises and accolades to Jesus, as we thank him for allowing the victory. Meanwhile, redirecting all fault to what may be, and rarely ever is, the Devil. This part of our culture creates adult Black Men who simply lack ability to own their own shit. While relinquishing a great deal of pride, that is absolutely instrumental in developing a healthy self confidence. A confidence that is required to achieve the victorious success that validate our purposes for being here. Pro-creating, while certainly a Human obligation. Perfecting the act of doing so, should not be the only source for the Black Man to feel pridefull.

Part of the reason why we are fighting so hard at Griffin Cleft to sustain social events that require are patrons to leave their homes, and electronics behind. Is so, that within the positive atmosphere created for their socializing and entertaining needs, to be met with much exhileration. Satisfying exhileration, that provokes thought and conversation that will lead us to identifying much of the systemic design. Design, that is cleverly woven into lives that we perceive as our own choices. When the harsh reality begs to question, do we really choose?

God has taken the burden of Sin, among much more that could debilitate us. So that we can flurish in the purposes that cement the legacies we leave behind. Legacies, that add to the stories of our historic accomplishment. Which all reside within the abundant resource that should, in fact, begin the feeling that is pridefullness in our Black Men.

Black Men. Know that each day you gain in your life, is a day where mistakes can be both corrected and avoided. As far as "it" goes. You will never get "it" done. Therefore you can never get "it" right, nor wrong.

Just make that efforts which can catapult you to the dreams that you've been told are impossible, to make come true.

Tomorrow I have a lot of "Sugar Honey Iced Tea" to figure out. We begin filming Mad Science in just a couple of weeks. And, so far I have a script for a docu-reality show, about how my soulmate found me here on Earth. And, that together we would be guided by our Heavenly Father through the launch of Griffin Cleft.

Although, separation seems to be our best option. I have done a lot of pre-production work that includes storylines for the two of us. Not to mention our contractual agreement, and job responsibilities. I thought that AC would be the answer to my prayers, asking God for help to build this company. After building Three companies from scratch. This time around, I just don't have that youthful gumption to go at it alone. And, against all odds.

None the less. I'm delusional if I think that AC will make it through the challenges that are to come up ahead.

What I've decided to do for us today. Is take the back seat as assitant to AC. As effort to help him plan his test shoot this week. I know, right? Genius idea. Wish that I could say that I came up with it. Remeber, that was the excuse he gave me for making me cancel my plans with Mom, only for us to have still not been over to pick up AC's things.

Day Three of him living here with me, against my request to verify his current living situation. I think that since she is one of his regular models that has hired him before, this could be a real booking to prep for. I've seen her in his book. And, I did listen in on the call she and AC had earlier today.

Of course the first thing we needed to do was shop. I took him to some of the local shopping areas in Harlem. I was impressed with his ability to negotiate terms of consignment with the design houses. He was very confident in his knowledge and industry lingo. After we finished photographing all of the possible choices he'd prepare for his consultation. But, via the local Kennedy's to get a bag of greasy fried chicken, honey! We stroll along on one of the prettiest days this summer has offered. Just, chopping it up and eating our chicken a la cart. However.

The wind becomes so fierce, that there are moments we turn and walk backwards. So that we can breath. The temperature doesn't feel any cooler. They're no bottles and plastic shopping bags spinning in the air. Amomng the other characteristics that ordinarilly accompaning winds such as these. Nothing, looked out of the ordinanary. But, somehow. As if we were in an isolated vortex, traveling through another universe identical in appearance. Only, two sets of weather.

He notices a pair of stilletos, at a street vendor on East 138th street, that aren't in such bad shape. Just beneathe the Beige platformed stilleto. He gasps at the stunning elogance in a pair of Black Satin heels, studded with short gold spikes. They only appeared to be about Three inches. "I normally like to put higher heels on a girl Amber's build. But, something about these Black heels are really speaking to me." He tells me, as I debate the amount of budget that I will allot for his shoes. Ten Bucks later, we continue moving forward, towards home. Heels in hand.

As AC prepped some billing and scheduling. I took the liberty of creating his pitch. I staged Two editorial shots for the heels that we had gotten back. Since, we did not actually bring any of the clothes back with us. I could not add much more to the generic, snap shots taken at their showrooms. However, for the shoes. I really let that aristic beast come out to convey a great story in the photos. Complete with sets, lighting, and post production work. Paired with my little photo essay with some of the most cleverly worded, fasion forward, consulting that I'd ever done before.

About Ninty minutes after I began, he was back to stressing. "I really need her to chose the looks from the Designers who wont charge me rental. That way, I can make more money." Oh boy, here we go again. I just reply with, "Check your in box."

Why would he include designers who would charge a rental if he couldn't afford it? I've made the mistake of spoiling him. But, now I have to put my feet down. And, make him stick with the budget that I gave him. Definitely holding on to the purse strings, now that I've become aware that my investment is just going back home, to live with Mama. For now my role is assistant. That's exactly all he is going to get from me, today.

The major difference bewteen AC and I, is that I don't try to figure anything out. Other than the moments I get stuck in trying to figure anything out. But, ordinarilly. I just ask God, "Ok Dad, this was your idea. What do you want me to do?" My relationship with God is one where I never care more than he does about things he wants me to do. I've found that when he needs something done, it's easiest to just submit to him like a robot and just let the mother board control my actions.

Once I am free to go back to doing what I want, I feel better knowing that I have invested some sweat equity, into my Heavenly inheritance. I don't ever know when I am getting it right. I just know that I can't get it wrong. Neither can AC. But, AC has not only the obsessive fear of being wrong. Also, he allows the enemy to present a resulting punishment. A little too easily for my tatse.

If you are a believer in a greater power that we could agree upon, for argument's sake, would be, "God." Even if the answer is simply because you can't wrap your mind around how we got here. Then, why do you only seek advice for the emergincies needing miraculous sollutions? Have you ever asked him what he felt about you? Does he have a plan specifically for you? What is his evaluation of Human Evollution, and. Are we missing anything that he just wishes we adopt into our societies?

I've been in chill mode for so long that I forgot he was there, year after year. As I did nothing much more than bullshit. God reminds me that he was forced to witness it, while missing the days of how I use to depend on him everyday. And, include him in or staff meetings and major career decisions. So now to hear the word, "Chanelling" thrown around to describe what one of my gifts is, feels less active than the act of simple observation.

What God wants us to understand is that Death does not equate end. There is only one way into Earth and one way out. For humans anyway. But, for the spirit of Humans. The spirit of energy. The spirits of all that is, we transistion.

Yet, the depth of their exitence is only as deep as your belief.

I can hear AC on the phone with his Model in the background. It sounds like she is confirming her delight for the design of her test shoot. And, we can now move forward with the rest. I create a terms sheet with contract points and other contractual information for AC to send over to her agency. Turns out, an agency that represents both AC and the Model he's styling this Friday. The next day I get a call from said agent.

I hate having to be part of a circle, inclusive of mild professionalism from entrepreneurs. God and I spoke at length of what this next and continuing chapters of my professional life must become. With my other companies, back in Los Angeles. I was just like the Agent. I raised a great deal of money. Met minimum feduciary criteria. But, when I was in the arena of tip top competitors, my green colors would shine psychodelic.

I began my first company, ONG Entertainment at only Nineteen years old. I planned the entertainment for Large Scale Branding, and Demo research events. My client list included companies such as Nike, Activision, and Salah of Beverly Hills.

My most impressive selling point was that I was so young. Most anyone that met me hadn't ever met a young man holding it down in the way that I did. Especially with no help from Mommy or Daddy. However, almost reaching the over the hill age for gays to be put to pasture at the ripe old age of Forty. No one is going to overlook small imperfections and blame it on my age. God has made it very clear that the expectations will be much greater. Like, uniformity.

Once we have produced a corporate process for any individual goal. The process is the same whenever that goal needs to be obtained in expansion. I can't just leave pertanent details such as accreditation, NDA for resources and intellectual property to AC. I'm just finding out that he even has an Agent. However, the Agent has not invested the time nor the money that I have into forwarding AC's career. I'm not even doing it for a return equal to the return that an Agent gets sitting by the phone. AC's success only adds to the value of Griffin Cleft, Mad Science, and my personal brand. Where would Brad be if Rachel Zoe had bombed? Collecting Government Assistance like me I suppose.

So the Agent decides to put the shoot on hold, until he can secure another stylist. AC has been fired because his agent can't seem to find understanding in my involvement. While I am on the phone with the Agent, AC is trying desperately to call the Model. But, of all times his phone has locked him out. On the other hand, my call waiting has just rung the model through to my phone. Thinking she was calling the office for AC, I was surprised when she asked to speak to me.

She explained to me that her Agent tried finding information of my company on the web. He could not, and felt as though I was someone trying to take advantage of his client. It's amazing that in this town, that is the first conclusion that everyone assumes. I could list the reasons of how any professional would have recognized game. So I won't apologize for operating in a high efficiency of professionalism. Even before I have made one public acknowldegement of our existance. With the discount AC is offering, simply to build his roster of clients and contractors. He desrves some other honorable compensation. Since our terms were unusual, we needed them in writing for sure.

Before I finished explaining the rest. She was already on our side. She was ready to put this issue to bed, but she had to do more convincing for their agent. What strange hold does this guy have on them? I've had plenty of agents in my life, and I've always been under the assumption that I pay them to work for me. I'm viewed as a threat because he can not validate the full scope of Griffin Cleft. That's the problem with introducing new processes. Until, the processes have been validated, rewarded, and duplicated expanssively. It confuses those not in it. But, that's the point. Why would I need to create a corporate anything that mirrors something already in exitence.

So now we wait. I didn't get to my mother's house again last night. Because, we're still trying to get over to his current "home," to get his things. Until we do, I'd rather not leave him here alone. And, I can't take him to my Mom's, because he seems to travel with dark clouds above him. Although she and I both live in NY, my sisters can drive to her faster than, the time it takes me to reach her, on public transit. The trip takes more than Two and a Half hours. So, I have subtracted a couple of weeks off of my weelkly visits since AC has been here.

While I wait, I decide to continue doing some interior design work to the bathroom, and adjacent areas. We make very little small talk as we share the space and lapsing time, that I will never regain. I'm always in fear of the guilt that my Mom will die, because I told her I would be there and didn't. And, then I have to live with the guilt of letting her last memory of me be a disappointing one. As I paint, I hear screaming, and the trowing of papers and books accross the room.

Frustrated because his agent has just decided to drop him. And, he is still locked out of his phone. I tell him to walk to the mobile store, a few blocks away. They'll reset his phone. I'd like to go back to painting. Perhaps, turn up some Scarlatti while I luxurate in the glow of the Yellow, Eggshell paint I zig zag across my bathroom walls. But, here I am stuck, again. Stuck of course pacing back and forth, because he is walking out side. As I pace, the observer that becomes Mark begins constructing characteristics and idiosyncrencies that, if isloated, wouldn't raise an alarm.

But, culmatively. I have to say that. Blue AC, his nickname to differinciate him from the air conditioner. It was Blue AC because, he only buys a Blue Bic Lighter from the store. Not, because he is always singing the Blues. But, now I wonder.

The Spirit of Truth then let it dawn upon me, that this entire time we had been fellowshiping. He has been leading me through God's knowing through Christianity. But, I assumed because we were so similar, that he didn't need to be lead through God's knowing of the Law of Attraction. It was time he got his first taste, of Esther Hicks.

I am in the belief that everything is of God. Well, I was....I think. Maybe I still am. I will have to let you know how that's going when I write the Blue Book. Now let's say that it is. That would infact mean, the multitude of options, variations and outcomes you can use to manipulate your purpose in God's Grand Design. They'd be infinite. Abraham is Infinite Wisdom. Philosophised by Esther Hicks with a single basic understanding. The universe is a drum. What you bang on it, bangs right back at you. It bangs in the same vibrating energy in which you bang your drum. Vibrations that mold your life and provide the evidences of The Law of Attraction.

My favorite lesson has to do with exploring all the options the begat happiness. If Christianity is your primary source of happiness, and for some reason you can't tap into it. Try something else untill God taps you on the shoulder and says, "Sorry about that, Yo. What had happened was...." And, trust me. He can tell you plenty of ways you could have done it better if you were him. But, since we're not. We can relax in the comfort that his word says that we should not worship any Gods before him. I promise you he couldn't give a fuck about what you did after you put him first.

Until Jesus gives me more clarity to the Bible, I'm old school in disecting every odd phrasing, that's filled with duplicity. With my webster and my Vatican/parochial web sites. Where those fail, I turn to Esther Hicks. And, AC was about to attract his introduction, to Abraham.

7

When AC comes back from resetting his phone. I thought I would ride the coat tails of that bit of happy news, and introduce him to Abraham. But, I'd forgotten that his momentary resitance to happiness would not allow him to see the positive evidence, resetting his phone would introduce. So as he returned, he was already on One Hundred. Going on about how the reset deleted his contacts and that he'd not backed them up any where. Worst of all, he didn't know his Mother's number by heart. Who doesn't know their Mother's number Heart? Or at least have it posted on their refrigerator. Or, something? Ugh!

More great evidence from the mobile technology age. You never have to remeber numbers by heart any more. So unable to tear him away from his phone again, as he submits to the power of Facebook. That place where everyone can spend hours reliving today's tragedy, and pointing out what they hate about each other. Kind of like lust, I'd rather flee. Most would call me crazy when they see me talking to myself. I just get so much more from talking to God, outside of what I gain from talking to strangers online.

So I scrap the Abraham introduction, and attempt to shift his focus to getting his things. More excuses, like he couldn't get a hold of his landlord. There is no way I am canceling on Mom again. So I have to take him with me. I decide to have him wait in a coffee shop, nearby. While I would go check on Mom and my Uncle. But, when it was time to go, he was too tired and didn't understand why I didn't want to just leave him at home.

That was just it. It wasn't his home. We've managed to overstep many boundaries, that I tried very hard to establish when we began. Throughout all of this, would you believe that I still had to fight off his sexual advances. I don't know why I did, but I chose that time to tell him I was letting him go. I was confused as to whether I was letting him go, or if he'd already quit. He hadn't really done any work this week, plus he said he was going home to Georgia this weekend.

When it was time to leave for my Mother's. He sat there. Like a child throwing a tantrum, he could not understand why he would be loosing his job? Feeling as though I was chained to my apartment, I begged him to continue this pity party on the road. The road toward my Mom's. I call Mom, and cancel yet again. She prays with me after I tell her about what's going on. Answers from her prayer led me to chose this evening to make his introduction to Abraham.

Whenever I make the introduction of Abraham to a gay man. I usually use the clip of someone asking what Abraham thinks of Gay Marriage. Abraham is brilliantly eloquent in how she describes everything. The ability to gain new perspective on adversity, is most surprising in watching her clips. She paints the picture of a Tug of War where on one side Homosexuals and the other hetrosexuals. This single statement is arugued from both sides. "I want you to change yourself, so that I can feel better about myself." The fight only ends when one side gives in. Otherwise the struggle continues.

AC wants me to change so that he can feel better about himself. And, I just want him to leave. Odd that I love him, or at least have love for him some how. Yet, all I want is for him to leave. Everything about our relationship seems wrong. He couldn't keep the act up. He has changed faster than a new wife, after nuptuals have been exchanged. Refusing to leave, but also refusing to be in the same room as me. Often running to the bathroom in tears. What the hell am I supposed to do?

Everytime he returns from the bathroom, he's suddenly heard the voice of God. How special? All of a sudden he's chosen. Yet, his life is full of destruction and he's as conflicted as they come. I'm not sure what whispers he hears. But, my God doesn't whisper. His voice echoes strong and through the clouds. And, offers me clarity, and missing links, and messages that gives me joy and prosperity.

I needed to know what I was dealing with so I looked up the origin of his full name. Everything that came up in my search was dark and demonic. Evidence of how people through out history with this name had destroyed worlds and were liasons to Evil.

Terrified that AC had brought with him, the devil to my place. I wanted to get him out of my place. So we went for a walk. There is a sky mural over my kitchen. So often, you feel as though you've been outside when you've only stood beneath it. God, often has to remind me that I am not out side when speaking to him in the kitchen. I'm not sure why it matters, but I do get clarity from our walks.

When we walked outside, I felt so much better. Yet AC did not. It was like his mnd could only comprehend the opposite of what I meant when I spoke. I would say, "I just want you to get your life together." And, he would here "I'm never going to be with someone who's life is not together." Do you know what it feels like to have two separate conversations as one?

As he stood to loose so much. His focus was loosing the husband he had found in me, when that was never an option for him. At least outside, he didn't seem to be controled by something. In the apartment, the pupils of his eyes would shift from big black dots, to tiny snake-like dots. While he made this look on his face that seemed like he was trying to induce mind control.

"Could this really be happening?" I would think to myself. He'd tell me his view on how I treated him, and then make that face as to telecpathically force my mind to view the situation his way. It would never work. I stood up for myself in ways that I wouldn't have ever imagined that I would. Including, threatening to call the police on him if he could not control himself. I've never witnessed someone pleed their case in such an aggressive manor before. Giving all new meaning to the phrase, throwing himself at me.

He convinces me to let him stay the night and he would leave after the photo shoot tomorrow. With all of that going on, I slept like a baby. I was in such a happy place that nothing could ruin it. Law of attraction was working in my favor, and God had chosen me specifically. But, when I wake. I learn that AC had never been to bed. He claimed to be searching for a place to live. He had been waiting for a call to view an apartment this morning, but could not answer his phone. The screen wouldn't recognize his touch or something.

He mentioned that he had the chance to discover the Abraham-Hicks web site. Which is why he gained the motivation to stay up all night searching for an apartment. He seemed to be in a much better place. Even brushing off the first couple of phone mishaps as the equivalent of God telling him that those places weren't for him. Sometimes Man's rejection can be God's protection.

Him taking the iniative to give it a real try to Stay in New York, and working on his career was quite impressive. I've witnessed before how quickly Abraham can turn someone's life around. Problem is, that the positive jolt is short lived the first time around. Staying positive One Hundred percent of the time is unnatural and requires so much effort, that many of us can't continue to ignore evidence of negativity. The act of living a positive life is in fact living a lifetime of ignoring all that is negative.

I was hoping that his first spout would last at least long enough to get him out of my house. As for most Millenials, locking him out of his phone becomes the fastest way to test his strength. And he fails. I don't know why so many Americans buy Japanese phones and expect them to work flawlessly over here. But, as he persists I continue in God's orders for me to assist him the hell up out my house. I try hard to ignore that drama coming from the corner where AC hovers over a locked screen trying desperately to unlock it.

I start packing up diva's things for her shoot today. With each item I stow away into a purple tote, I see my investments going into the trash. Drop, drop, and drop, until we're all packed up. Just as I was about to scurry into the kitchen and burn up some grub. My phone starts to ring. I'd imagine God would want me to be the Assistant Caterer too. So, it's only befitting for me to put on my Craft Service Apron. Yes honey. A bitch can change them Aprons and Hats. Some how I knew that this was my day to show off. What's to debate? All the decisions had been made.

"Praise The Lord!" That's how my Grandmother, Bishop Viola Griffin, would answer the phone. "Yes hello, this is Amber. I'm trying to reach AC and he called me from this number yesterday." This is what I hear on the onther line. "It's for you, pumpkin." Don't take that to heart. Pumpkin is usually a pet name I use for sarcasm. Like when you know something is about to go down. But, you've learned to stay positive and you want nothing more than to ignore someone else's drama. Well, in this case.

I could tell that "Amber" was calling to bring him some bad news. My feelings were validated as I saw a grown, Six Feet Six, Black ass man fall to his knees in tears. I've been there witnessing his Month long dilema. But, come on God! He's trying.

It's hard for me to offer advice to people in bad times. My advice is usually steeped within the philosphies taught by Abraham. Which is usually the last thing people want to hear. Advice for the bad, from someone who doesn't believe in bad. AC's ex-agent was determined that his models would not work with him. So when "Amber" arrives in New York from Boston. Where she resides. The agent met her at Penn Station, with an entire new crew. Besides apologies, there was nothing for "Amber" to offer AC.

He immediately falls to the conclusion that with out the money earned from working the shoot, he could not afford to move or go back to Georgia. I'm not sure of how long God wants me to be his little helper. But, I think it's time I exercise my free will. So first I call Ms. "Amber" back. If I couldn't convince her to hire AC, I could at least threaten her to pay a cancellation fee.

Although Amber sounds like a lovely woman. I think that she in the agent played the hell out of AC. My hands are tied. I don't know what New York's laws on squating are. But, I just want him out of here. So I go outside for a walk to clear my head, for the new plans that will replace the clutter.

So while outside I get a call from a stalker I have, who lives in Boston. I thought that he'd forgotten about me, but here he is after not hearing from him in a couple of years. When I amswer, it's the usual. I'm going to kick your ass and such. So I hang up on him. But, not before he could warn me about visiting Boston, where he now lives.

We actually use to date, back when I was Fourteen years old. I asked his father to sponsor me, so that I could attend Ballet Camp. But, he told his son that I offered him sex for money. If you saw his father, you'd know how untrue that was. But, here we are decades later and his father has not fest up, and he continues to find me here and there. Just to taunt me over the phone.

Getting the call from the Boston stalker gave me the worst feelings. Just pulled me right out of my cloud of happiness. Then I remembered that I left my keys at home. I wondered if AC could be stupid enough to lock me out of my own house. So I ran back at top speed. When I turned onto my block, I see AC leaving my building. I zip down the street, screaming at the top of my lungs. "AC! AC, wait!" When I approach the front of my building, I notice that the man who I thought was AC. Was not.

I punch in my front door code, which rings directly to my phone. But, now guess who gets locked out of his phone. Yup, me! I stand there awaiting the next person to leave my building, so that I can go in. Praying that I am not locked out of my apartment. I fly up the stairs to open the door, noticing that AC probably never even left the Bathroom. Maybe I'm over reacting. It's just something is so strange.

My phone keeps ringing, but because I now recognize the Boston Area code. I ignore the calls assuming it's Mr. stalker. Just when I start to coax AC out of the Bathroom I get a text from "Amber" saying that she needed to speak to AC. She'd been one of the missed calls that I ignored. Yes! From Boston. Now after speaking to Boston Stalker, I'm a little thrown by "Amber" being from there. AC calls her back, but gets her voice mail. "I thought you said she just texted you?" AC asks me. "She did." I reply. "Why are you lying to me? I just called her and she did not answer." He says to me, with that eye transition, as if he's reading my mind for the truth.

I wish you could see that evil eye look. It makes me laugh out loud, which only adds fuel to the flames. Saved by the bell. She's calling his phone now. Only, AC can't swipe his screen to answer the call. Locked out again I suppose. When his voice mail finally answers, he is allowed in his phone, and then dials "Amber" back. Not even Ten seconds into the call, and my phone starts to ring. If "Amber" is calling me, who the fuck is AC talking to? When she asks me for AC, I go next to the bathroom door to listen in on AC's conversation.

I again try to get "Amber" to commit to some kind of cancellation fee, so AC could get some kind of money out this ordeal. As she answers my questions, I can hear AC on the otherside of the bathroom speaking in sync with "Amber." Could it be that there is some app that exists to transpose human voices to sound like other voices? Is this some sort of Magic?

These are the questions raging through my mind. Then my mind quickly shifts to my Mother. I haven't seen her in weeks! If he can do this with Amber, how do I know that he hasn't had my mother murdered, and when I call her. His app answers on the other line. Abraham says that anything that you can imagine, must exist. I pursue investigation immiediately. Before drawing any imaginative conclusions.

I hang up and call my mom. Of course it goes straight to voice mail. I'm not sure that I would even believe that it was her on the otherline, if she did answer. I have to go to Staten Island.

First, I got to get AC out of my house. I let him know that I'm on to him. And, that I believe that he has orchestrated the whole thing. I'm not sure if it was just to waste my time, or to have a place to live. I joke with him, "Maybe it's just to keep me away from my Mom." Then he gives the strangest reply. He pokes his cheeks out with that evil eye, and says "I am going to get to your Mother!"

Delivered with such disdain and courage. Puffy chest and swiveled hips. "What the fuck do you want with my Mother?" You know Bueens are Momma's Boys. AC is about to make me grab my can opener. He claims that I heard him wrong, yet doesn't remember what he was trying to say. As if it wasn't him speaking. Whatever the case is. It's time for AC to leave. As I walk him out with his things, I explain to him that I will call the cops if he knocks on my door again.

As he leaves, I stay outside. It's a beautiful day. Sun is shinning along with the sounds of birds chirping and children's laughter blowing in the subdude winds. I decided to wait since I didn't feel that it would be over that easily. Rest assured. The wind began to pick up. Trash aroud me began to spiral, as clouds began swirling at high speeds. I couldn't see him yet, but I knew that AC was closely approaching. The wind became so strong that I had to grab on to a mail box in order to stay put. Once he turns the corner on to East 141st Street, I grab my phone and dial 911.

He turned around assuming, I suppose, that I am calling the Police on him. When in fact I was calling for my Mom. I asked police to go check on my Mom and Uncle. By the time I hang up, the wind has retured to it's mild flow. I go up stairs, and as I enter my apratment my phone rings with a call from my Mom alerting me that the Police had just been to her house. I'm glad that they're alright. But, I have to get my locks changed before I go see my Mom with my own eyes.

It doesn't take long for you know who to show up after AC leaves. "The Talls won." I hear God tell me. "What the hell are you talking about? More importantly, where the hell have you been?" Don't be alarmed, this is how I always talk to God. We've known each other for so long that pleasantries of worship are thrown out of the window. We get right down to business. "Most of what you understand is opposite of the truth. Soon, you'll have clarity when you get to the other side of things." With the day I've had, I have no time for God's riddles.

It's been a while since I spoke to my older Sister. So, instead of debating philosophy with Pops in Heaven, I give her a call. After I finsih giving her the low down on all the drama that I've delt with since speaking to her last. She told me that I got caught up in spiritual warfare.

When I get off the phone, you know that I have to conduct my own research. Wikipedia describes Spiritual Warfare as a Christian stand against preternatural evil forces. As I study more on the subject, I suddenlly feel unsafe in my own home. As though I need to burn some sage or something. I then hear in a female's voice "Don't let another day come in with the Devil in your house."

My sister had shared with me some prayers to pray when in spiritual warfare. I begin praying them. I walk around my apartment, plus around my floor praying these words against evil forces. When I finish, I find myself standing in the kitchen underneath that painted sky. I hear from the same female voice, "You only have Seven minutes left." I assume that the female voice is the Spirit of Truth, so I am taking what I hear from her to be the truth. I'm not sure what to do after noticing that it will be midnight in seven minutes. finally she says, "You're not really outside."

This time it's as though my body had been stalling, like a car, with some one attacking a car ignition, desperately trying to get it to crank up. But, instead of igniting an engine. She, was trying to ignite my inner GPS. I feel as though the painted sky above my kitchen has tricked my body into thinking that I am outside, where I am supposed to be at that moment. When really I am inside negating my Heavenly duties. Once my engine starts, my legs start moving me around the house. Grabbing items like keys, phone, and a Jacket. I rush down the stairs and out of my building.

The feel of relief comes over me as I learn that I got outside with Four minutes to spare. Not sure of why, exactly. The feeling of accomplishment rushes throughout my body. I live on the NE corner of a one way "T" where my street disects East 141st Street. I cross over East 141st street to the SE corner to get a closer look at the street sign. As I read 1, 4, and then 1, I remeber God telling me something in a dream about me being chosen for Eye For an Eye. Not sure how, but I know they're connected. At midnight I began reciting the Lord's Prayer aloud. Again, my body's GPS has taken over, and knows the ritual that I am greenly participating in.

About Thirty seconds after I finished the Lord's Prayer, I look up to the sky. Taking off from the Laguardia Airport is some kind of space craft. It's not of equal measurments to a Human, but. It's shape is designed like one. The wings looked like two extended arms, and the tail looked like two legs. Keep in mind that it's as large as a plane, however. It looks, if you are familar with, *Buzz Light* Year from Disney's "The Toy Story".

It's flying North and West as it flies over my building. Then it circles to fly South and West back past where I was witnessing all of this. Down on East 141st street. I hear the plane laugh out a psycotic laughing sound, as if he had just won. Okay, am I going crazy? Had I just esxperienced pcycosis? "Congratulations, you just sent the Devil home." God whispers in my ear.

Like a lunatic, I call everyone in my family to let them know that I had just sent the Devil home. I say lunatic because it was nearly One in the AM. As if I were drunk dialing. Only, I had nothing to alter my state of mind at all. Completly sober. Maybe a little drunk on life. I'm telling you, the idea of confessing Christ as my Lord and Savior and only a week later God chooses me. Then I ace a week long series of exams, tests, and challenges. Finally, to know that I single handidly sent the Devil home, saving us all on Earth from his rath. I'm feeling like the luckiest Nathan Detroit in the world.

Not sure if it was because I was visited in a dream, or what. But, when I woke I knew someone had died. But, AC still holds on to my power of remebering my dreams. So frustrating that I know, but I dont. You know what I mean? Immediately I call everyone I assume it could be. Last on the list, still left to call would be AC of course. AC appears in my apartment.

Well, his voice only, "I'm going to kill myself." I hear him say. I jump on the phone and call him but when he answers. My phone hangs up. When he calls me, his phone disconnects. We repeat this pattern a few more times before God tries to explain.

"You guys can't do the same things at the same times any more." God says. I don't have time for bullshit, so I ignore him as I am trying to save my Soulmmate's life. All of a sudden all of the electricity goes out in my apartment. Then it comes on again. It goes out, it comes on. I start to cry hysterically as I yell out obscenities at the top of my Lungs. "You guys can't do the same things at the same times any more." I hear God repeat to me. "What are you talking about?" I ask him. "If you take a breath in, he can not. If you are asleep, he must remain awake. If his hands are wet, your hands are dry." My Father explains.

My phone rings, my front door registers on the caller ID. When I answer it is AC on the other end. Not paying attention to that, "Don't kill yourself!" I yell out. "How did you know?" He's already crying as he asks me. "You were here, and told me." I tell him. "Wait, are you down stairs?" I ask him. "No, I am at home." He answers. I didn't let him know what my caller ID registered. I had a feeling that God was at work. Allowing me to get the ease of mind that I was desperately seeking. I'm not sure if he assumed that I would handle the truth differently or what. But, here I am talking to my soul, I think.

The rest of the conversation continues with loving goodbyes. We were both clear that we would never see each other again. Here's the tripiest part. Because I woke up inside the apartment, and he was in the world. I thought that when God told me the "Talls" won, he meant that AC got out. And I was held captive in the soul. I really thought that we had switched places some how. He made it to Earth and I am now somewhere that looks like Earth. Yet, I am the only one there. As I look out my window, there is not a person or car on the street. Have not stopped crying yet, accept to yell out louder cries when I realize that I will never see my family again.

I thought that I was taken, because I didn't make it back to my Mother's house. I couldn't work any of my electronics to contact anyone. I am asking God why he has played such a cruel joke on me. He offered me this unexpected reply.

"Imagine if I had to assign one letter of the English Alphabet to each of my chosen. That would leave me with only Twenty Six Subjects. With all that must be done in the World to ensure it's continued existence. I would have to give each of my chosen a purpose, which only allows for us to accomplish Twenty Six goals, at once. Because you work insynctively with me. There would never be a reason for you to work together. Through me, you know one another. You love one another. You respect and honor one another. You are the witnesses of how the World moves. Your timming is ordered in a way that every second of everyday is witnessed for me. You are the validators of the Truths of the World. You are the decidors of why the World should still exist. Which requires each of your purposes to remain un-united."

8 JESUS TAKE THE WHEEL

As I turn onto Mom's block, I am excited to see her. Not only because I want to check in on her. But, this would be the first time she has seen me since I became a True Christian. I can't wait for her to quiz me. If any one knows the word of God, it's my Mom. She has read the Bible cover to cover mutiple times. I've watched her fall asllep with the Bible in her hands only to see it right back there when I wake up the next morning.

When I arrive in front of the house. My Uncle is outside already. "Hey Mark. You know the Cops were here yesterday to check on us." "Yea, I was scared because I couldn't reach you guys." I tell him. I'm not quite sure why, but something seems very different about Uncle Gordon. He's much more lucid today. As if his mental illness has disappeared. The screen door squeaks open as my Mom appears. That door squeaks loud enough for the entire block to hear when you open it. She doesn't want to get it fixed, because it alarms her to people entering her house.

As I give her a hug and enter the living room. I can feel a strange energy. You know how it feels to enter a room where everyone has been talking about you, and they stop once you've entered it? It felt knid of like that. Like she was expecting me. It seemed like my Uncle and Mom were preparing some kind of ritual to acknowledge the new level of faith that I had gained. She starts by rubbing bless oil on my forehead while reciting prayer. For those of you unfamiliar with bless oil. It's basically olive oil that a clergyman has prayed over and that God has blessed.

When we all sit, my uncle immiediately falls asleep. My Mom tells me that we'll have plenty of time tomorrow for Bible study. But, she wants me to stay the night, and that I should get some sleep. When she exits the room, my Uncle pops awake. He offers me some tomato juice, just as my Mom enters. "No, drink the cranberry juice instead." She tells me. I poor the crandberry juice and sit across from Uncle Gordon. He gives me a strange look and shakes his head no. I look down at the juice that I have placed on the floor at my feet. I look back up at his strange look, and I could tell that he was signaling for me not to drink the juice.

I'm sitting there confused when my Mom enters the livingroom, where we're sitting. "Why haven't you drank your juice yet?" "I'll drink it Mom." I tell her. My uncle gets up, walks into the kitchen and stands next to the refridgerator in my view. So I get up and follow him.

Not sure why I am suddenlly taking cues from the man who couldn't tell you today's date. But, he seems to be the only one inside familiar with this ritual. Keep in mind that both my Uncle and Mom were raised by Bishop Griffin. God already told me that there is so much more to Christianity than average people know. That combined, summed up that I was not the authority.

I wondered if My uncle had gone through similar processes of being chosen when he was my age. I also wondered if it is the cause of his skitzofrenia, and whether I'd become that way too. I leave the cranberry juice on the floor so that my Mom thinks that I am drinking it. But, while in the kitchen I poor the cup of tomato juice, and guzzle it back like a Bloody Mary shot. My uncle and I go back to the living room where he begins to tell me stories from his youth. Stories of his involvement in my Grandmother's church. He jumps up and runs outside when my mother enters the livingroom.

May sound strange, but I am noticing that, the Three of us can't stay in the same room at the same time. I wonder why, but something is just not adding up. I thought, perhaps my Mom may have spiked one of the juices with my Uncle's medicine. She sometimes does that since she can't always count on him to take his medicine. While my Mom goes to the bathroom, my uncle comes back inside. "Don't you have to throw up?" He asks me. I just kind of look at him. You know the look when someone is trying to get the right answer out of you. That's me right now. Full of confusion. "Sarah, I think Mark is sick." He tells my Mom.

"What's wrong with you baby?" My mom asks. I'm not quite sure how to answer that just yet. Keep in mind that I feel perfectly fine. I look over to Old Man Gordon to see if he can provide me with an answer. The look in his eyes is enough for me to say, "Yea Mom. I'm not feeling well." "You got a fever?" She asks me. "I think he needs to throw up." My Uncle says wisely as he exits out of the house. My Mom goes to get me some antacid, as I sit on the couch. Sitting, and wondering where God is in all of this. I ask God what is going on and I hear, "You better go throw that shit up." I'm not sure if it were God or my Uncle channeling me from outside. But, I go to do as I'm told.

As I make my way to the bathroom, my Mom calls me to her bedroom. "Do you have any money?" She asks me. "Yes, how much do you need?" I reply. "I owe your uncle Forty bucks." I give her the Forty bucks, and she goes to the livingroom to give the money to my Uncle. As I exit the bedroom My uncle stops me, "Hey Mark, you need some money?" As he hands me the Forty dollars that I had just given him, he grabs my hand and nudges me towards the bathroom. As I am in the bathroom, I hear my mother ask, "Where did he go?"

As I stand at the toilet conflicted about my duty, I figure I should share with them the confusion that is taking over me. When I open the door, I see my mother and uncle standing there. My Mom is holding Two pair of scissors that she quickly tries hiding behind her back. After I open the door, and have already seen them. "What's wrong honey?" My mom asks. "Don't say anything." I hear God tell me. "Nothing Mom. Hold on." I tell them as I go back into the bathroom and shut the door. Back to my conflicted latrine. I proceed with the duty at hand.

Reminding me of my Ballet School Years, when I was Bolemic. I stick my finger down my throat, gagging myself to the point of volmitting. When I am done, the toilet looks like a crime scene. Red tomato juice everywhere. Noticing the bloody scene in front of me, it dawned on me that the red juices could have represented the blood of Jesus. If that were the case, I think I may have just fucked up by regurgitating his blood from my body. To fix, the problem I took one of the Twenties and threw it into the toilet.

I figured if I were to flush away the blood of Jesus. I should show that I could flush money down as well. Proving that I am not manipulated by any Earthly posession. Blood or Money.

Now onto the problem of what my Mom intended to do with those scissors. I was stilll a little consumed by the energy of tragedy or death. I thought that my Uncle was going to kill my Mom. Like she knew that in order for me to be chosen, she had to be sacraficed. Perhaps I watch too much T.V.

My Mom bursts through the bathroom door. "Mark honey. Mark baby. Oh my God!" She yells after looking at the bloody tomato juice all over the bathroom. "I'm sorry Mom, I'll clean it up." "No. You come lay down and I'll clean it up." She says. As I am laying in bed, "Mom I love you so much." I look over at my Uncle who has now returned back to Crazy Man Gordon.

I didn't arrive tired yet somehow, I fell straight off to sleep. Right as my Mom releases me from the hug we shared. Not before wondering a little bit of nonsense first. God told me that if I died, I would wake up next to him. So, I wondered if my Mom and my Uncle had poisoned me, so that I could find out the truth in that. Perhaps there bodies were equiped with GPS systems similar to mine. And, that they were under control as much as I was. I wont keep you in suspense. I wake up the next morning.

Sometimes God has a hard time of convincing me that his gifts are actually gifts. After I fell to sleep, he comes to me in a Dream. He assures me that I am not dead. But, my life wouldn't be free of death. He didn't tell me how soon, but he told me that it would be soon. One of my loved ones would be dying soon.

He said that I came into the Earth with Two spirits. My right side was made up of the Spirit of Christ. The left side was made up of the Spirit of Satan. "Now that you have learned how to exorcise Spirits from within you. You can no longer be the host for Satan's Spirit. But, you were not designed to live with only one spirit. You need at least one other, for validation. Until I replace it with the death of your loved one's spirit. Since there is no validation that you live. You are now half dead."

Crispy Bacon. Rich coffee. Sausage, muffins, melted cheese. These are the aromas that brought me back to life. As I opened my eyes, I see my mother leaning over me. "Good morning, honey. Come on and wash up for breakfast." She tells me. "Mom, where is Uncle Gordon?" "I don't know." She answers, as I burst into tears. "What is wrong with you?" She asks me. "Uncle Gordon is inside of me." I screech out. "What the hell are you talking about? Gordon is outside somewhere." She replies. "No, he's dead. We've been chosen! God had to make him part of me." As she pulls her hand back to slap me out of dillerium, The loud squeak of the screen door pulls her focus. I turn and scream, "Ahhhhh! You're alive." They just crack up at me like they were being entertained by Whoopi on the view.

"I'm glad that you can find me so funny." This is too much, I need a cigarette. I wasn't much of a smoker before this point. But, when you are chosen by God, the insanity of his mysterious ways can drive you absolutely crazy. So leave Obama alone. Some men are Chosen to smoke.

As I've formentioned, there is something weird in me. The weirdest of all is feeling though I'm loosing it. It's kind of like if you, for some reason, have to land an airplane that is already in mid air. And, you see red lights flashing, and hear the beeps and bells of emergency warnings. But, you're a doctor. You have no idea how to fly a plane. Nor do you understand how to stabilize any of the malfunctioning equipment. That's how it feels. Especially when it comes to spirits and religon.

I tend to lean toward the more spiritual side. However, until this point in life. I always assumed that I was never alone, just never had a way to validate it. But, now through the teachings of my Christ Ancestory, I have gained more validation than I needed to have. I know that my writing can never cover the true since of confusion and fear that comes along, with this Chosen process. I just don't have much of a mind that remebers the negative outcomes, nor emotions that accompany them.

Anyone who knows me, thinks that I have the worse memory of anyone. And, that I have probably smoked too much weed. But I kid you not. I'm like a toddler. You know how you can give them a spanking and five minutes later they are smiling again? I'm still on the fence as to whether or not that's a good thing. Sometimes in Los Angeles, Jeremiah and I would go out on the town. And, run into someone that we've known. In my inebriated state, I'll run up to them with love and hugs. And as we returned home. Jeremiah would often whisper to me, "You know that you don't like that girl?"

I would always respond the same way, "Really? Why?" Then, Jeremiah would have to continue with the story of how someone, at once upon time gave me the perception that they've wronged me. But, what I dwell on when I've been wronged, is my participation in it. Desperately trying to identify and correct the part of me that could allow such tragic events to happen to me. So that's what I often take away from being "Played." I release all blame from the offender, while I own my own shit, as Iyanla would say.

As I am out front on my Mom's porch smoking a life saving, cool Neport. The Holy spirit told me to google the definition of AC's name again. As I do, I find that the search results were completly different. I could not find a single thing on the whole of the internet to relate to what I had found in my results, just the day before. I mean, miraculous praise and accolades for many people thrught history who carried the same name. The meaning itself had changed to become more positive. Nothing about demons or tragic death, like before.

Quite confused about it all, and God ain't touching it. He's already moved on to today's agenda. He's told me to get dressed to go for my morning run. Yes, he's very concerend with my health and appearance. It may sound obsurd, but he even guides me through recipies that I have been cooking or baking for years. The nerve! Like a parent looking over their child's shoulder the first time the child cooks an egg, for themselves.

There he is. In my ear, reminding me of the shit that I already know very well. I must admit, there are plenty of times where he has actually talked me through variations on my own recipies that improve the taste of my food tremendously. Even still, I just don't know why he cares so much.

Ear buds in, work out clothes on. The joy of living in the same city as my Mom for a change, comes with the perk of having a closet full of clothes at her house. So, I can be dressed for whatever occasion. Even when I don't plan ahead for said occasion. The GPS has taken over once again, and I have no reason for why I have chosen the route to run. I suppose that I didn't choose. I would never run so far from my Mom's house just to work out. I would probably circle the block of her neighborhood. But, ofcourse this wouldn't be any ordinary run.

Today he has me paying special attention to the cars, not only on the road. But the cars also parked, and how they assoicate with the adresses that they are parked in front of.

For example, I would be on Dryer Avenue, in fron ot 123 Dryer Avenue. And, then he'd pull my focus to the Toyota Camry in the drive way. I would take that to mean that, Japan is currently in first place. I may have failed to metion sooner that within the spiritual warfare games you represent your country. It's just like the Olympics. I was currently holding the Fifth place solidly for America. I bet for most of us American's, we'd probably say that I was loosing. But, he reiderates that the battle is not mine. Nor yours to loose, but it is the Lords. That helps me not to feel so unsuccessful in my Fifth place position at the moment.

I wonder often about the others chosen. I would imagine that Asians would do well on any tests. But, I wonder what Japanese kid is running around Tokyo participating with the same level of energy that I am. The Japanese seem to be more logical and their actions and thoughts than God's process allows for. God was able to find the One young man as disconnected with status quo as I am? In other, words stupid as fuck?

My run leads me to Walgreens. God has me purchase over Two Hundred dollars worth of things that I will need in the coming weeks. He often will give me reasoning for the actions I must make. But, lack the understanding of why. Hey had me buy a pack of "Puppy Liners." I don't have a dog. Yet, when I am having a weekend full of sloppy sex, they come in handy to protect floors and funiture.

However, his purpose was for my nephews who had just gotten a dog. "Pumpkin Bone," the dog's name. Can't seem to stop pissing all over my younger sister's house. Which means that his days are numbered. God wants the boys to have the responsibility of having to take care of somthing who's lively hood depends soley on their commitment to it. So here was my participation to prevent my sis from getting rid of "Pumpkin Bone."

He also made me get plant and flower seeds. His task is that I am to take the oldest one out with the dog first. He would be responisble for walking the dog in the mornings. While on their morning walks, the oldest nephew would plant the flower seeds. Then, when the younger nephew walks "Pumpkin Bone" after school, he'd see the flowers bloom.

The younger nephew would be more motivated to take "Pumpkin Bone" for walks. He'd not only be walking, but also checking to see what new sprouts of life would be waiting for him. And then, of course to use the pads in the house until the family got into the rhythm of caring for their new addition.

I have become so obedient, that spending Two Hundred bucks on shit that I can't see a need for at the moment doesn't phase me one bit. There once was a time, where a trip like this would have taken me hours to complete. I would not have only questioned my needs for the items. But, also whether I have the money to make such extravagant purchases. Yes! I have become so poor, that at times shopping at Wallgreens is extravagant. Can you believe it? Me? The Ballet dancer who has toured around the world. Performed for Kings and Presidents. Not to mention, walked away from a very comfortable life, that I managed to build in Los Angeles.

The rain is pouring, and it's only raining water, unfortunately... well maybe not too unfortunate. Considerng the drout that our West Coast countrymen are experiencing. As I am returning home, Uncle Gordon pops out of the woods, near my Mom's house. It's the second time I've seen him on my run. I better stop and acknowledge him this time. So odd how he appears to be perfectly normal today.

As my Uncle Gordon talks to his imaginery Air Force buddies. That keep him stuck in the Sixties or Seventies, you can hear all kinds of obsurd profanity and senseless smut. Yet, somehow in his conscious state of mind. He is the most judgemental, God fearing and confessing person that you'll meet.

He still feels that woman shouldn't wear pants. Especially, if she's not in the kitchen doing hosework barefoot. Uncle Gordon is not looking so well. I suppose, playing his part in this spiritual warfare is kicking his ass, the same way that it is kicking mine. There seems to be very little that is spiritual about the warfare. Yet, instead it is physically manifested games, challenges, and judgement that takes a great toll on your mental and physical strength.

Would you believe that as Uncle Gordon and I sit at a bus stop sharing a cigarette. I still feel like he has died, and my Mom just doesn't know it yet. Yes, here I am looking into the eys of my living Uncle. But, some how my mind will not process the fact that he has not died. Nor, anyone else for that matter. I still give AC a call to find out if he has jumped off a bridge yet. But, no. by the time I reach him. He has jumped on a bus, instead. A one way ticket to Georgia is in his posession, and he has already boarded.

How's that for closure? After he had been there to get me through the spiritual warfare that God has chosen me to soldier in. And, I msut tell you. He was excellent. I wish that I had enough pages to tell you minute for minute, play by play. Just know this. AC is truly a believer. His encouragement, and companionship during a very confusing period may have been the total sum of my success. Or, he could be the total blame for why America is not currently holding First place.

Whatever the case, it is a bitter sweet memory. No matter how bitter, I would not exchange our time together for all the riches in the world. I pray that he finds the resolve that he is seeking. And, will make his way aback to the Fashion capitol of the world. To continue moving in the career of styling that I am convinced is his purpose. Come on, he styled me literally in his sleep.

Also happening today is my return to the Church. For so long, I hated the idea of it all together. But, today I am also returning to my home church, Christian Cultural Center. Led by my favorite teacher ever, Pastor A. R. Bernard. It's the only church in New York City that I have ever known. When I visit churches with traditional "Black" preachers. And you all know the stereotype that I am referring to. It's hard for me to take them serious.

Once I visited my younger sister's church, with a pastor that terrifies me. Not for me, but for the pastor's own sake. She was preaching so aggressively with explosive trembling, and fierce vibratto of her voice. I was more focused on the posibility that she could have a heart attack, more than I focused on the message she delivered.

Pastor Bernard, is not that way at all. It's why he calls his church a Cultural Center instead of a church. He has really branded the feel of an institution that study's where we came from and where we are heading. Every message is taught with the charming distinction of an Ivy Leauge, education. When it's time to give our tithes and offering. He'll even joke that business men all over the world pay him a lot of money to get the counseling that he give us for free every week.

I am not a fan of Mega Churches. That's all I have been in for my entire life. Joined one when I moved to Los Angeles. And, when I would visit Jeremiah's church with his family. His was equally as Mega. I sometimes wish I could actually touch my spiritual advisor. But, as most of them do. Pastor Bernard rolls with the security detail worthy of protecting presidents. I wish I could share with him of how his teachings continue to bless my life. Messages like, anything done to the extreme, abuse is enevidable.

That was a message that he taught when I was only Twelve years old. How ever, it has been the one saving grace that has kept me from addiction. As an artist, there are many opportunities to develop addictive, and self destructive behaviors we're prone to attract.

As a a Libra. Making the decision to always do the right thing. Includes a process as indicisive as every other decision that we Libra tend to debate into procrastination. So it becomes a never ending battle that is unfortunately mine. And, not the the Lords.

When I arrive at church. My energy is on One Thousand. Which oddly enough can be very off putting for the Jaded everyday New Yorker. In the one location to let your happiness explode from within your core of happiness. I was told by the book store's cashier, to relax. I guess I may have looked like a crackhead. Or, at least seemed like one, rather.

How dare she? I'm dressed and groomed for a visit to the White House. I'm that Bueen that you see in the Choir Stands dressed for the nines. And, the only man with professionally styled, salon hair. Evrything on me in its place.

But, I don't allow the church staff's low energy, to provoke a change in my happy being. I understand achieving the level of happiness that I often failed, when attempted by the company I keep. As I enter the Sactuary of CCC, I see that Pastor Bernard has clearly been chosing.

While in the middle of his sermon, he says. "He's here? I better hurry up." Speaking to himself, out of character. It could have been a coincident. But, I read it as his respons to God. And, that God had already told him that I was coming to receive the word of God from him. And, Pastor Bernard had a special message for me.

Just as I take my seat, in the upstairs balcony's section with couches desined for the new Mothers, to tend to their babies. Which I have never sat in, BTW. I sat alone. Well as alone as I can ever be. A feeling of being in the propper section came over me. Since Pator Bernard was aware that I would be there today, I need to sit where he could not see me. Remember, God's chosen really shouldn't ever be in the same relm. Let alone in the same building.

I can't get my pad and pen out fast enough. I arrived at the perfect time to get the base of my studies for the week. He begins to talk about the origin of the calendar as we know it today. And, the origin of what it use to exist as. All the information was centered around the Samritans. And, the mathmatics developed by the Samaritans that not only shape our world. But, in my opinion could actually support theories of time travel and Quantum Mechanics.

And, just like that. Five minutes later. He was done. I couldn't believe it. After all the excitment and drama to get me to that physical location. I had only made it in time for the last Five minutes, of the last of Three services that day. As I almost relax into the perception that I had missed out on the whole of the knowing. God whispers to me, "What is for you. Is for you."

Without missing a beat, I move on. Continuing to bask in my happy Glory and Grace. With the exception of a couple minutes of crying. But, I expect that these were tears of Joy. I'm not sure of why, really.

We experience death of people close to us all the time. However, the fact that my nephew, Marcus, was only Nine years old. Coupled with my intentions to work on our relationship. And, the help that I would eventually offer my older sister once I got my shit together. Really allowed me to fully grasp the concept of tomorrow not being promised. When he died, I was kind of just chilling.

Catching up on all the years of Television that I had missed due to my grind, seemed to be more important to me than actually stablizing my well being and career. Althoug I am still young enough to continue in my dance career, I was clear when leaving California that I was hangging up my Pointe Shoes. Dance, however marvelous at first. Included too many moments of degrading situations. Days where I dreded going to rehearsals because I hated the material that I would be rehearsing.

Since it was always my goal to become a choreographer. I have always looked at the choregraphy that was taught to me to be lacking in artistry. Especially when compared to my own. You'd be surprised at how many gigs dancers just show up to, making up their own steps. While the choreographer, who is supposed to be our leader. Just sits back and collects a check and the credit for what we dancers should have been paid and accredited for. None the less, I hadn't given much thought to the timeline of how long my transition would take.

I was clear that I wanted to explore the creation of artistic products within the For-Profit sectors of Entertainment Production. I find that conglomerate control, by the big commercial entities of our world prevent choreographers form creating true works of art. However, I am in the belief that art doesn't have to be thrown out due to commercialism. Just look at movies like Mulan Rouge, Chicago, and other inovative works create by companies such as the Metropolitan Opera.

I also knew eactly which projects I would begin with. How I would present them. I even had scripts completed for a T.V. Series, Three Musical Stage Plays, and for the work out series that I would package as scripted films. So within all of my Grand schematics. I had not considered when to shift my focus to them. Or, to actually take the roles and title of an Entertainment Producer.

But, now. Here we are. Here we go! God has shown up and chosen me. Chosen to show me the truth and what I already believed to be true. Yet, you really understand how it's only the faith of a mustard seed once he shows up. I have made many actions deeply rooted in Christian Idiology. Some that were greatly beneficial, and some that lacked success. However, I never stopped believing that God is real. That his son is Jesus Christ. And, that I am also a child of God.

No matter how much of a faggot that my spiritual leaders may think I am.

My knowing, trust, faith, and belief was not only minor. It lacked the senseful abilities and understandings that he tells us right in his word. It's just when you hear it all told to you in a different manor. Detached from doctrine and politics. You can go back to his written word understanding the truest validity, and purposes for what and why his word is written.

I had actually never shopped at the CCC gift shop before. But, here I am now. All grown up, and actually doing my grown man thing, with cash to burn. As I leave CCC and walk leisurely to the train station to go home. Brookly, and especially the delapadated flatlands that CCC is in. Looked especially picturesque today. Skies specifically painted for me, with streets adoring beautiful murals, artwork, and manicured store fronts that didn't exist at points before.

I am physically and emotionally drained. The extremeties of my emotion being pushed to all opposing limits. I just can't wait to get home. It should be to catch up on sleep. But, I had just gotten really relative answers to guide me on my studies. So I am running home now, to commence indepth study about the Samaritans.

Once again. Home alone. Sitting in the dark quietly, and in silence. I had made it home after having a pretty good day with church. For the most part. Yet, as I sit in what felt like an empty apartment. I guess Luther was right, when he sang a house is not a home. I was excited to continue studying when I left church. Continuing to learn God's process as well as the facts as we know it from Bible study. But, once I got home I couldn't break away from the over analyzation of my own thoughts.

I can get that God has shown up to take credit for the ideas, strategies, and products that I have been living with for the past Four years. Finding evidence that validates that I have been chosen for something that comes with great reward, well in the moment. Doesn't seem so lucky. Trust me, the event in which God shows you himself, as sort of apprecation for your faith in him is its own reward. But, when I think back on the things he told me in the begining. I was expecting the Glory from being chosen by God.

However, as I look around. It seems as though there is very little Glory to go around. If I'm half dead. Can I get half of all my heavenly inheritance. I'd love to see half of the heavenly riches that were promised to me.

I know how much of a spoiled brat I must sound like. Since an imrpoved life has been prepped, given to me, and is ready for use. I can only thank him for that. But, if I knew that AC could be some random stranger that God has borrowed for the purposes of training and communicating with me. I would have ignored everything that seemed real about our relationship. I certainly would not have invested so much money in him. Nor, would I have included him in my travel plans. Or schedule.

As I sit back in the emptiness that lacks the positive energy and memories that my soulmmate brought to this house. I ping pong back and forth in my head. Wondering what the hell was that all about. A very ellaborate process that yielded very little results. And in the end, I am still left holding the reigns alone.

Carving out my piece of the New World has been an expensive journey that has resulted with events that leaves me in question. I question, is this the Devil playing tricks on me? Or, has God really chosen to unveil some of his mystique to me? At any rate, the fact that I now believe in the existence of the Devil, my own belief system has been rocked.

I always assumed that he was fiction created by the church to scare non believers into becoming Christians. Knowing this, can I really continue traveling around in the world ignoring that both bad and good exists?

The thoughts behind my beliefs that there is no bad, comes from the idea that we will never get it done. Life here on earth and ever expanding has never stopped, and will never. As we transition through the fullness of being, it's an everlasting journey. With a level head, and clever manipulation. You can turn the events of a bad situation into positive, good outcome.

With that understanding I was able to release judgement, doubt, guilt and many other characteristics that derive from concluding what is bad. As if anything ever concludes. Like the craziness of this process.

Remeber me telling you that I felt that my deceased nephew, Marcus, had been controlling my phone. Well as I sit, I get a text on that BlackBerry phone. That's the phone with the issues under Marcus' control. And, the phone that I had recently disconnected. So you can imagine my surprise to get a text out of the blue, on it.

I don't recognize the number. But, the text reads "This is Marcus and Kent. I came for you. Diana is not a believer." Diana is my younger sister. Kent is the son of Ms. Four Octives out in California. Kent was also born with Autism. Since I've heard that people with Autism don't live long lives, I thought that Kent had died. Why else would Kent be with Marcus? But, once I read the last line of his text, "Don't let my Mom ride in Diana's car." My worry shifted to my biological family.

Living in the belief that Marcus has been staying with me since he transitioned. It was not hard for me to believe that some how he had figured out a way to send me a text message. But, still! What was I supposed to do? Text him back. So I toss the phone onto the bed, and return to my pontification of the problems at hand.

Before I could ease into my chair comfortably. I get another text. This time it reads, "It's Jesus." Coming in from the same number that Marcus had just texted me from.

So tired of all the games, and now this? Is this some kind of joke? When I see the text I become a little angry. I don't believe that it is either of them texting me. And, assume some asshole is playing a sick joke that includes disrespect for my family, in using Marcus to manipulate me. As the phone hits the bed, I get another text. It reads, "It's me! God! Shit!" When I see that Jesus has just cursed me out I think, who could know that I curse God out? I'd never told anyone before writing this book about that part of our relationship.

He continues in the next text with, "Today is my Birthday, they have it wrong. Every Four years I am allowed to show myself to Nine people. But, this year I have chosen you to be a Tenth person."

There is no more convincing that needs to be done. I'm on board. Here we go, Jesus show me yourself. But, I wait by the phone and no other texts come through. One hour. Then two hours. I'm looking out of the window, and watching shadows in my apartment. I didn't know what he meant by show himself. I thought that since I heard his father's voice. Maybe his trick would be to appear and I could see him. You know how they always show him in movies, with a halo and gold light above his head?

When nothing happend. When I did not get any more texts. I called my older sister to explain to her of how Marcus had been communicating with me from beyond. As spiritually intuitive as she is, I knew that she would believe me. Her response instead was, "Let that boy rest in peace. What you are doing is demonic."

I tell her that I would come to Maryland on the weekend to show her my proof. She tells me that she was not going to be in town this weekend, because she was going to California the next day. When I asked what time her flight left, she told me "I'm not flying, I am driving with Diana."

What! I don't want to alarm her over the phone. But, I knew that if Marcus has asked me to prevent his mother from getting into a car with Diana. He must have known the danger in it. Everyone in my family knows that Diana is the most dangerous driver in it. I was once with her as she needed to exit the freeway. She missed the exit, but instead of driving safely to the next exit and making a U-turn. Diana drives over the apron and through the grass at full speed like she was driving a getaway car being chased by the LAPD.

I beg her not to leave until I get there. I couldn't let her know over the phone. But, she needs to read this message from her son before she jumps in the car that might take her to her death. As much as I would love to sit here waiting on Jesus to show himself to me, I feel obligated to protect my family. So, I run down to Penn Station to get the first train smoking to Baltimore.

9 BLACK LIVES MATTER

"Get up." I was in the middle of a dead sleep when I hear them whisper. I say them, even though I hear a single voice. I don't hear other voices in the same way that I hear God's. Most of the other voices are so hollow, as if they're coming from shadows. So, until I can gain momentum twoard clarity. I call everyone who isn't the voice of God, They.

They continue as my GPS boots up, energizing my body to full errection. "The Girls have a headstart on us. They're on their way, heading West, without you." I assume by Girls, they mean my Two sisters who are driving cross country as we speak. "Get your things and meet me outside. Your dress attire has been pulled and hung for you. Wear comfortable walking shoes." They advise.

Remember, that I was to be in Los Angeles by now, for my "Play, Play" Sister's Juneteenth Gala. Yes, it's coming back to you now. Mrs. Four Octives. I received a text earlier from Marcus. Exactly, from beyond the grave. That's why I pretty much ignored it. I don't know what to make of You're in Kent's World? It also mentioned something from Jesus telling me that he's let the twins out for a week to stir up trouble. Every Fourth Year, after Jesus has allowed his chosen to know him in truth. He releases the Gemini Twins. I suppose in my story, Marcus and Kent are the Gemini Twins.

A little worried about Kent, since Marcus seems to be up to mischive. I wonder how Kent's health is doing. It's too late to call her now, but I must remember to give the Four Octive Family a call first thing in the morning, to make sure that Kent is not the death that I am expecting. I should probably send some flowers and a note canceling my reservation for the Junteenth Event.

I've been putting off a visit back home so long now, that I would hate to go now. Before, I have some grand announcement that meets the expectancy of my fans. Griffin Cleft is already about a year behind schedule. And, God, the Spiritual Warfare and now Jesus are all unanticipated, yet welcomed deviations from my original plan and timeline.

None the less, I am outside. On the block, at Four in the morning. All the stores and business offices are closed, and dimly lit. Yet, the streets are full of pedestrian traffic. I have no clue what I am doing. So I kind of just follow along with the other people on the block participating in 2015's Spiritual Warfare Games.

Again, we're divided by country of origin and religious faith. Often, I tend to be the only American Black person that resides in my neighborhood. And, clearly tonight's ritual and exercise was introduced long before Lutheran, Methodist and Non-denominational doctrine. So, I watch the Mexican Catholics very carefully as we disect the walking ritual patterns.

We start on the North East corner of East 141st street. We recite the Lord's prayer, and wait for the nearest traffic light in view to turn green so that we may walk. The main goal of the exercise is to bind on Earth, what is bound in Heaven in terms of land equity. I will definitely elaborate more in boks to come of details, not only within the Spiritual Warfare Games. But, also on the scientific inclusion that is much to complicated for me to lay out in only one short chapter.

I am clear that the Purple Period is the only period that I will play the Spiritual Warfare Games. However, this Purple book. Is not at all about Spiritual Warfare. I am only providing dots that you can connect in effort to maintane your sanity. As, I tick, tick, boom, away.

God has already explained to me that people will call me Crazy, among arrogant, and narsacistic. That is why you have to be a true believer in order to be chosen. Pause. Let me go back, and attempt to not sound so arrogant. You have to have a solid belief system in order to be chosen.

A great Pastor once said that, when God deals with you it's like having an expensive suit tailored. If you interupt the Tailor with a visit before the Tailor has completed the alterations, you'd yell, "What have you done to my expensive suit?" When in fact, the professional process of tailoring requires one to take apart the suit, and then reconstruct the alterations into the magic that makes your suit red carpet worthy.

So, as we begin to take apart what Mark knows to be true. And, introduce doubt, and confusion into the fullness of Mark's belief. Mark, has to have unfaltering moments where he can stand in his truth. Validated, proven, scientific succession that comprises undisputable truth in which Mark may rest in, as Mark's World is being created.

It's funny, however. For someone that stands in his truth as deliberately as I do. I can't seem to move as quickly as everyone else needs me to. Creation, and twisted resolution is all that excites them in the process. Otherwise, known as Evollution!

Or, retrograde in tonight's case. Jesus and I are one in the same, in that we'd prefer a phone call. Opposed to texting. I take it he's had enough. Especially, when I only take them serious enough to witness. But, never do I reply. Preserving a little of my own sanity here. I'm sane enough to know that modern technology allows Them to communicate clarity to us. But, responding to their clarity only takes acknowledgement. Therefore, Jesus decides to start whispering in my ear. Meanwhile, Marcus still texts.

Jeremiah and I have a saying, "In your ear." When people offer you warnings, and then you reach the deciding moment. You, don't really have a choice in the matter, because the warning is already in your ear. That's why I hate when my Mom gives me the nightly lists of worries that I should take with me on my travels. My Mom is often the worse voice to have in my Ear. Coversely, Jeremiah's Mom, which I call Ma West, is often the better motherly voice to have in my ear. However, to be fearless in our creation, we don't ever really need either motherlly voice in our ears.

"Make a list of all the things that remind you of your Mother." This, Jesus speaks to me, as we stroll leisurly through the neighborhood. Jesus doesn't sound as sexy as his father to me. He has an Owen Wilson, kind of tone and approach to delivering information to me. He feel's like a very cool, old and familiar friend. When we chat, I get him and he gets me. We have such an easy breezy relationship. To start, that is.

Jesus loves to ask you to put your turst in him. Then, blame you for every disappointing outcome of your hard work, faith, and strength. We've both made several mistakes along the way. However, I musn't forget that my life was perfectly fine before he and his father arrived. The hardest moments of my struggle were over, and I was actively rebuiding the quality of life that I enjoy most. Freedoms to do and go as I please.

Obligation, commitment, and responsibility were things that seemed foreign upon first hearing the words. As God, puts it. Much of the work that we've done in 2014 and 2015, is in effort to fix my lack of interest in intimate commitment. That's how he cleverly words my fear of intimacy issues.

My last partner broke up with me in 2012. Although, it would be the shortest relationship of my life, only lasting Eight months. Ending it was more devistating than any heartbreak that I'd ever known.

I know that I have loved other men in the past. But, I would almost venture to say that loving Tyler was the first time I'd known True Love. Entering the relationship, I knew with our travel schedules being so hectic. There was no possible way that we could be monogomous. So, we agreed that it would be an open relationship, when we were in cities on opposite ends of the world.

Funny, enough though. I had no interest in sleeping with anyone besides Tyler. And, would hate to learn when Tyler would be advantages in our agreement. Go figure, right? It was mostly my idea.

Monogamy has never been something that I required. Although I honor it, when it was required of me. My longest relationship lasted only Three years. And, the others were at least Two years, each. So, for that short of a time, I could understand monogamy. With as busy as I can get, years typically fly by. But, I still don't get the forever part. My Heavenly Father calling upon me to now choose a husband, is sadder than it is encouraging news.

I love unique characteristics about almost all ethnicities. So, how do you choose just one? Understandably, a Black Man would be my first choice. Not, just in appearance. But, with parents and ancestors rooted within the United States. Carribean, Latin, and African men all come with cultural differences that, I feel would be hard to manage within a life long commitment.

"Each day, you must decide Four aggreements that you will honor. They can be the exact same agreements each day." Jesus tells me, as I cross over, South and west, then around an eletric pole. Entering the street between only American makes of cars, finally connecting the edges of what we map out for the Christians. I know. Confusing? I'd probably be more confused if my neighborhood didn't already have the Catholic Street Signs such as St. Mary's Place, Wales Street, and others important to our world.

Jesus continues, "Men should begin each day with physical, laboring, work. However, on select days. You must schedule earlier wake times. So, that you can get in household chores, repairs, and organising done, before moving forward with your work day." Jesus clearly has the objective to offer the clarity, and missing links to add steam to Griffin Cleft's embarkment.

As I dosedo around, and throughout the neighborhood. I find that Jesus has led me to the farthest Northeast that I have been in New York. It is the place where Queens borders the Bronx. Most people doesn't even know that it exists. A few steps West, and I am in Harlem. A few steps in the opposite direction will land me in Queens. Bridges make it very convenient to connect with these Outer Burough neighborhoods. Yet, stil if I tell someone that I live in the Bronx. They'll usually persist with the laundry list of reasons why The Bronx is too far to travel.

It will be one of my tasks to introduce the new wonders of the South Bronx to the rest of New York. You can really feel the fight of its residents to improve the quality of life here in what is for now, Mark's world. God, Jesus, and the Spirit of truth continues to reiderate to me that I. Along, with Their help, will create an entirely new world. The status quo of it will be exactly what I and the citizens of Mark's world validate it to become.

For the first time in my life, I notice that the Moon is rising. It is an absolutely stunning, Quarter moon piercing the gorgeous blue sky as it breaks a part clouds and lets the sun in. Could this be possible. Both, the Sun and Moon rising simultaneously. The moon is traveling at such a high speed, as if somone has pressed the fast forward button on the remote. A remote control that conducts all the sky's activity.

You should really spend an afternoon just observing how much goes on up there. As Jesus and I continue our walk, and ritual, I give new meaning to the kid with his head stuck in the clouds. I can barely tear my eyes away long enough to watch for cars on the road. "Now that the Glory of the sky has been unlocked, you must always wear sunglasses. They will call you pretencious for wearing them at night, but you must even wear them then. Do not stare directly into the sun. She will take your eyes." Of course, as Jesus shares this with me, all I can do is look at the Sun. And then at the Moon. And, back at the Sun. "Stop it before you hurt yourself." Jesus says.

Like a kid discovering them for the first time. I have so many questions about why and how they exist. As I stir at the sun, I can see it disected into several circles of light. The circles continue moving inward, getting brighter towards the Sun's center. As the center turns white, the whiteness travels toward me. As if there is a special force that connects with that of whom I am to be. I continue to be alarmed at the speed in which the moon is rising. I question, to see if it is in fact real. "There is not any magic, or theatrics at work. Well, none other than my Father at work building your World. This Moon is yours, and it is new. That is why you can identify it's speed so easily. It is a Moon taking it's place in the Univers for the very first time.

You are not in the place you once new as Earth. And, where you currently exist is momentary. This is the Validating World. Here, you will be introduced to many ideas that you will have to decide upon and commit to. This evening you will be visited by the council of your first Secret Society to choose from. The council must choose and accept you as a valid candidate to join them. Each night this week, you will be vissited by a different council, and society." Jesus explains. "How will I know? What time? How will I communicate with them? " I ask him.

"The Spirit of Truth will be with you. That's how you will know. Just like on Earth, God has given you free will here. So, don't feel obligated to choose, due to the council's persuasive persual of you. I have chosen you. You are now chosen. I am the last to choose. Through me and your belief in me, you will live eternally. Fear nothing besides the Lord. Don't fall for false intents. Nor, false truths and empty promises. Plan your given. Plan in my name, and you will need not give anymore than you plan." Jesus, explains to me.

Still traveling through the ritual, Jesus calls my attention to various buildings, and happenings surrounding us. Near a Bridge that connects Queesn some how. And, underneath the new Quarter Moon. I see a line of Pick Up trucks heading East on the Street where we were, and then traveling up and over the bridge to Queens. The trucks were all blue. I could not tell what their specific purpose or contents were. However, it seemed as though they were adding an extension to the City New York.

Right before my eyes, buildings, brdges, and lands were being errected at record speeds. Expansive, maticulouslly calculated, construction that would extend New York's Easter Seaboard miles East of where it previously stood. "How are they able to work so quickly?" I ask Jesus. His reply. Are you ready for this? "You are on Mars. Martians have capabilities uncomprehentional to your Human capacity to comprehend."

Hearing something that outlandish, yields me to pause. I don't want to end up skitzofrenic, or worse. I desperately want to keep the faith that God, and Jesus have chosen me for something very special. But, how the hell did we go from Earth to Mars? Without me knowing or purposely making the journey.

"It will all make more since very soon. For now, stop trying to figure it out. Old World concepts and strategies will be used to identify possibilities. However, know that we are creating something that has never been created. All of your tasks will come with a duality that will often confuse you. I believe in you. My Father does not. He doesn't see why I would have the need to form a new world, nor does he think that a Human will ever be capable or strong enough to succeed at what must be done." We leave that little port city, to travel back west along East 138th Street.

The farther West I travel, I notice that the sky some how is seperating bright clouds from dark ones. The traffic has grown so thick, with cars holding travelers seeming to be making a great migration to the new world. But, if Mark's World is East. Why is everyone going West, and away from New York. A red light is born from the son, and traveling towards earth. It slowly descends, passing right in front of me and finally disappearing into the ground. With the ominus happenings both in the sky and here on the ground, I become overwhelmed.

"Relax. The changes you are witnessing is the exact evidence of your success and evidence that I have chosen well. Finally after Thousands of year. We can continue the evollution that provides the everlasting life that I have promised my True followers. The battle, all be it, not yours. Is certainly not over, and we've barely begun. You may return back to your base. After resting, you will be born again. Born new. Your immediate work will be the process of letting go. Be the cutting edge I have witnessed you to be, as you move forward in creation. Born free, safe, and with everlasting life."

Just as I turn onto Cypress Avenue, heading up to East 141st street, I witness the clouds partting to form the number Ten. I automatically assumed that I had just received Tens acrossed the Board. Tens for a job well done.

"No. The number Ten is movement. An erra where numbers began at Zero ends, and moving forward your numbers will begin at One." God tells me. I guess he had been witnessing what Jesus and I were up to. It still shocks me when he speaks. I never expect him to be present in my present.

God continues, "I have decided to do something special, to make up for you not being chosen originally. Your number is now Four. For the remainder of my process you will remain Four. I will be making you whole again shortly with the number Five. I will always be your number One. Together, the Three of us will complete the entire make up of Mark Griffin." Was I really not chosen? How can God forget about me. And forget that, my purpose was destined for the time in which I speak of?

"Your origin became faded as I adopted the belief that your original desinge would fail. But, you always knew. As your actions created confusion, that even I didn't think would bring you back to us. You've proven everyone wrong. You first entered Earth as Wolfgang Amadeus Mozart." God tells me. He then tells me who I was the Second time I existed as a Human on Earth. But, I didn't recognise the name, so I really don't remember who that Second being was. But, he continues. "In your Third Journey on Earth, you entered and left it as Josephine Baker. And, now your final days on Earth will be spent as Mark Griffin. You will no longer need to return to Earth. You can choose to retire with your family in Heaven. Or, continue to evolve farther away from me, and your Heavenly family."

I stay outside for a bit, before heading back into the apartment. It's such a gorgeous day out. And, there is beautiful artwork just above my head, as Leonardo Divinci has just manipulated the clouds to write the number Ten, clear as day. I mean with, big, bold, prominent articulation.

As I watch the neighbors walk past me. I'm amazed at how every single one of them is burried into their mobile devices. I am the only one, actually taking in what scenic inspiration matches with the feel of perfect, beautiful weather.

Out of nowhere, my skin begins to pop. If you have ever eaten those Pop Rock candies, that we all loved so much when we were kids. That's the sounds you hear all around me. This time, unlike when the Spirit of Truth first arrived. It's not all at once, and over quickly. It also doesn't feel like the same energy. It starts with a snap from my forehead first. Followed by a snap from my left arm. Then, here a snap, there a snap. Everywhere a snap, snap. What the fuck, yo!

I don't know if Global Warming is taking it's toll, or my anatomy is making adaptions to Mar's climate. Whatever the case, it's time for me to retire to base camp, and rest up for tonight.

When I get inside, The Spirit of Truth must think that it is night time. "No Honey. Just because my moon just rose, doesn't mean that it's the PM yet. I've been fucking around with Jesus and them since Four this morning. God is trying to work my ass to death. Making me whole dead. A bueen need his sleep." This is what I reply to her with, when she acts as if I'm scheduled to begin entertaining develish shit at Nine in the fucking morning!

When the Spirit of Truth is around, there is a constant flow of winds throughout my body that bring about goose bumps. You know that feeling that gives you goose bumps? It use to be a feeling I looked forward to experiencing. It has quickly grown to become the most annoying pain in my ass!

She thinks that if she just keeps swirling in and out of me, that I will give in and just play along with her. And, those of them that don't require sleep. I would have been ok, with my energy. If I hadn't only just arrived back in town from Baltimore. I arrived in NYC just after midnight, and got a couple of hours of sleep before Jesus woke me up.

Ok, enough already! I can't take her annoying me. I'm going to go over to my friend's place to sleep since my home is being made so uncomfortable, that I can no longer rest in it. I guess Jesus and his Father didn't work out the specifics of where they would live in Mark's World, before creating it. And, I feel often like my small One bedroom apartment/office is being over crowded with all the Christianpeeps, shacking up with me. I can't wait to kick them all out. But, for now to Cyril's place we go.

Cyril lives just over the Madison Avenue Bridge, and is a short walk for me. When I lived on West 137th Street, it was great to have a Bueen right around the corner from me. And, continually greatful at the present time to still have a Bueen within walking distance from me.

Just because an acquantance is both Black, and Gay, doesn't automatically make them a Bueen. Bueens, typically have their living situation so stable that. At any given time, night or day. I can ring his bell, and say, "Bueen, God is wearing my Black ass out! Let me just take a quick nap, and get a sip of grape Kool-Aid." We just roll like that.

I kid you not! I fell asleep as soon as my cheek hit that pillow. And, can you believe that these motherfuckers woke me up exactly Two hours after. I was barely in my comatose deep sleep zone. I don't know why, but for some reason. In my dreams, I can't seem to keep up with God. I find my way into Journies where I get lost. No clue as to where I should head. At the same time, I can't seem to remember how to get back to where I came from. So awake with no aswers, I see that they have found me.

I can't seem to escape them. When it's just God. He works the shit out of me, but. He speaks directly to me with precise articulation. There's not much left to be misinterpreted, or confused about. When it's just Jesus, and my little buddies, Marcus and Kent. I can't work my electronics, but we be good up in here. But, the others violate my personal space. Forcing my acknowledgement is an opression like you can't imagine. No matter how many times I ask them to stop touching me, they wont.

I can feel them moving throughout the house. And, now on the right side of my lower back, I feel a little zap of vibration. I think it's a way for them to stand in agreement with me. God gives me very little time to make adjustments, due to my culture shock.

The reaons why, he compares to, "Imagine that there was a clone of you, that we only powered up yesterday. You would be living the current life of what was cloned. Your brain will never comprehend, that your memories never happened to you. And, that you are actaully being re-taught to do all the things that you think you remember doing. But, that's only if you were a clone, that we only powered up yesterday."

My first Autodiadect education with God was "The Purpose of a Thing." My Pastor Bernard use to always say that you have to understand the Purpose of a Thing. When trying to succeed at or with that thing. When God introduced this year long study to me, back in 2012. He presented me with a single challenge. "Is there a thing that exists only because of one, single purpose." I can save you some time, by letting you know that you wont find one. There is one thing that I found, but. I'll save that for a better time.

I remember back a few years ago. When I produced a Ten episode pilot of a show we called Working Birls. It starred Jeremiah, my boyfirend at that time, and myself. The cameras followed us around, while living very full lives. Oh how I miss those days. We were all amazed at how well the finished product assembled. You only need One episode to pitch a show. But, I had no clue going into production. In the Old days, TV shows had pilot seasons. Not just One pilot episode.

None the less, here I go. Always doing the most. Killing ourselves with maticulous, Hollywood studio, kind of processes and strategies. Working our asses off to complete the most! But, in all of my producing I never stopped to consider one important question.

As production companies bgan to express their interest in picking up Working Birls. I met with several of the big time CEO's. Yes, girl! The producers of Housewives and everything else. And it was that meeting with the Housewives producers, that Brian Hale asked me. "Why do you want to be on Television?"

On the surface, the easy answer would be that I fell in love with Show Business at a very young age and have to exist in it. At all times. In whatever capacity. I am like Wendy in the way that I eat, shit, and breath entertainment. But, when I actually took the moment to consider the question. I could not produce the one answer that would define the progression of Working Birls, and myself as a producer.

I knew what the answer was. I tell you. I have this auto pilot memory that often needs to be lured, tricked, and caressed into working. My truest answer for most everything that I've done and will do in my life. Is to be the example, and guidance for the Young Bueens, growing up to replace me. To navigate through worlds, exploring everything possible. Yet, effected by none of it.

I'm often impressed by guys who can sound like a politician when debating politics. And, sound like an IT guy, when discussing technological advancedment. Not for employment purposes. Just to know shit, just because. In the moments where you really need to get someone to like you. Offerings that similate even small examples, relating to someones own culture. Could make or break the chemistry that allows you to move forward, in the budding relationship.

Every ounce of most that I can experience in life, is deliberately staged, or spoken with in direct relation to my Master's plan. I guess I've just been lucky that way. Even now, as I reconnect events that were particularly obscure in it's purposfull relation to me, at first. Have now gained greater understanding. Understanding of how each deliberate each moment and conversation of my life is.

Jesus is the first voice that I hear when I wake from my nearly comatose nap. He demands that I go back to our Base, and study the written word. He has assigned the book of Esther as my first Biblical Book, to study with deeper exploration. Ironically, a book named after one of my favorite people, Esther Hicks?

My 2013 study, taught by God. Is the study of words. I was on a date, shortly after I moved into West 137th Street. All be it bad manors to discuss business on a date. I was asked about the type of work that I am doing in this transition of one career to the next.

When I referred to Mad Science as kind of a reality show. My date was often confused, because he associated Reality Television to T.V. And, could not consider the idea of a web based T. V. series. So, he'd often interup my explanations, to solidify that we were talking about internet distribution whenever I would mention the words, Reality T.V.

As he repeated this manor of seeking extra clarafication on the plain old English vinacular that I was using. I noticed how specific he is in his speaking. And, also his relation to the origin of the words in which we spoke. So I say to him, in a laughing manor. "Words are very important to you." He replies to me by saying, "Words are important to everyone."

For the first time in my life. I centered face to face on the knowledge that I perhaps, may not speak the Queen's Eglish. I try to sound as educated as possible, most always. Especially when discussing and introducing business plans. But, here I am shown a new perception. One of where I am the hoodrat with a slang vocabulary of my own. A culture I try very hard not to participate in. Even when I reply to text messages or send e-mails. I will type the full spelling of words, that diagram the sentences, that convey my truest thoughts.

But, all of my practice of punctual grammer is of no consequence to me as he points out. That it could be my own wording that offers the confusion behind explaining the newness of our creations. Identifying Mad Science as a webisode, or. A web-based, realistic documentation of Griffin Cleft's taped production log. Is not only a mouth full. But, also includes language only just introduced into our Status Quo. Not really sure of how many people have adopted internet language into their everyday understandings.

Even though, I am clear that Mad Science is not a reality show at all. It seems to me to be the best generic label for what it is. But, perhaps I have found a new demographic, within my target audience. One, that can't compartamentalize ideas with T.V. included in the titling of the idea, that is to be distributed through the internet.

I bet he is the type of person that, when he tells you that he is at home, watching Mad Men. He would say, I am watching Mad Men on DVD. Just to avoid answering the question of, "Oh really, what channel is it on?" If you haven't yet, you should really watch a program, while on the phone with someone else who is watching the same program,and at the same time. It's loads of fun!

So as God and I pick apart the words within our English language. I become very fascinated with the origin of the word. I previously learned that anything that has gained a title, which includes it in our language. Must not only exist, but must of have existed once, purposely. If not currently purposful to us in present day. Often the fascination is derrived from the requirement of why the new word has been introduced.

Take for example the word, behavior. What would be the defining crisis in our history that would require the expressed identification of actions that would be commonly acceoted as behavior. I wonder if it were because of the actions of children or adults. Is it Bad behavior that garnered such a word. Was it some uncomprehentional character that must be extracted from mainstream, bugois societies?

Thakfully, the Book of Esther happens to be a quick read. Reading the Bible was never a chore that I ran to in happiness. Somehow, now knowing that there is much more to the frame work laid in the stories. It is becoming a fascinating, invigorating read. I believe that Jesus has drawn my attention to a specific fact in Esther. Not sure if it's mentioned only for the first time here, but one of the characters is a Munich. Even more fascinating since, currently on the Television each day. And, centered on Front page news is Kaitlyn Jenner's transformating sex change.

The time has arrived where I am to receive my first visitor (s). I learn that some of my heroes, currently positioned in the One percent, are helping out with my introduction to the Skull and Bones Society. Donald Trump, Warren Buffet, the Skull and Bones Society wastes no time jumping directly into my business side of things. Marcus Lemonis is the first to arrive. I'm sure his invitation included a smooth journey to me, since I had already been chaneling the name Marcus so heavily. Due, to me communicating with my nephew Marcus.

Still a little confused about how this all works. My first question has to do with the process in which we have established communication. Not quite sure if I hear his voice or if I am just channeling vibrations that equate with the sound of language I think I hear. But, I have to say, that I think that it is his voice that I am hearing. Shocked and amazed at God's brilliant magic that allows for miraculous wonder to inspire, I ask. "So do I just speak allowed as if we are having a private meeting? Or is it best for me to just communicate with you inside my mind?" Marcus then replies to me with, "Speak as if you would normally, as you describe the career goals you hope to achieve with Griffin Cleft."

I amaze myself in how easily adaptable I am able to roll with the mysterious ways that is God. When the time arrives for God to inrtoduce new to me. Or correct my perception of the ideas that are old. It most often feels like the next logical step. As opposed to some, huge, life alterating departure from the very essence that is me. But, a small variation in my own perception of my powerful capabilities. Thus providing small, circumstancial evidence that results in my increased belief. I'm telling you people. If you build it, they will come!

In 2014 the study was Cliches. After spending so much time with words and their origins. I noticed that, when you know the period associated with the creation of a word. You can find much evidence for what existed before you. Fifty years from now we can tell stories to our grandchildren. Of how we use to party and get "Turrnt Up." They might ask, "Why didn't you just call it being drunk?" Anyone who has ever been "Turrnt Up" before, knows that it is something all together different than being drunk.

Like Jesus promised each night I would receive my spiritual visitors. Some more bizzare than others. But, I survived it.

No one seems to be answering their phones in New York, so I decide to call some friends back West. Of course the first person I think of to call is Jeremiah. I'm from the era who spent hours talking on the phone every night. But, it's seems as though that we're a dying past time. Jeremiah doesn't answer my calls or texts still. Every few months I'll hit him up to see if he's had a change of heart yet.

Since he has not, I call Dawan. In Los Angeles everyone calls Dawan, Jeremiah, and I the Three Musketeers. At most Hollywood events you usually don't see one without the others. We've been doing our own things for a minute, and really hope to get the gang back together.

"Bueen." That's how we greet each other. "Hey Dawan." We catch each other up, on recent adventures in our efforts to take over the world, before he fills me in on the inner circle gossip. You think Hot Topics is hot. The shit that Dawan done heard on the street? Is the shit you want know.

The things that I am hearing about Jeremiah from Dawan are apauling. Not only is he ruining his own life, but trying to take me down with him. I'm working hard to put myself back together again. And, Jeremiah is continuing his family's tradition of Publicly flogging my reputation. I've been commissioned to write a musical play based on the life of a mutual friend of ours. Jeremiah is at the man's home with Dawan, and talking about me like a dog. The things I heard him call me were unbelievable!

Dawan had to be lying. I know that Bueen is not speaking to me right now. But, he has to still love me enough to not bash me in situations that could loose us money. He knows that we support our Moms and that taking food out of our mouths, is also like taking food from theirs.

I called the guy that I am writing a play about, who in formed me that it was true. "I wouldn't say bashing you, but. He certainly didn't have anything nice to say about you." Is what he told me.

I guess I shouldn't be surprised. But, as I stand over this frying chicken. I'm wondering Two things. Jeremiah must not ever be planning to forgive me. And, secondly. Did I use too much Seasoning Salt on this chicken?

I'm thinking about my Mom a lot lately. I am hearing things like, she's not my Mom. I even heard that my older Sister might be my real Mom. I don't know why I hear so many lies. But, I am figuring out that there mostly lies.

Jesus tells me it's because I can hear conversations that aren't meant for me to hear.

He and his Father visit me on Saturdays. The schedule that I follow is not a time or daily schedule equal to one that you would follow. While building Mark's world. We don't rest, so Griffin Cleft doesn't have a Sabath day on its calendar. Instead we have a second Friday.

This week I have been channeling with all of the Secret Councils, and now we must discuss my choice. It's a hard decision for me since each society has very different interests and controls. Of course I want to be rich, so Skull and Bones would be my first choice. But, that would require me to go to college which I am so against. Not the institutions, but the curriclum based academics and fee structures.

Yet Jesus has shared with me that he had me change my name to Mark. Because, Mark was originally chosen to tell the story of Jesus Christ's life as he lived and died. And, I would tell the story of his latest walk. In that case, Illuminati seemed to be the better fit. But, then I am naturally doing the things that the Son of the Knights Templar would be doing.

So, like choosing a husband, I don't understand how anyone settles on just one. But, since Jay-Z and Beyonce are my ideal romance. I choose Illuminati.

As soon as I choose, the house lights dim in the movie theater of my mind. The curtain rises on a young Mark Griffin, named Omhmar at the time. At age Seven, he is tagging along with his older Brother. Spending the day just visiting his homeboys, making runs, and eating out. Which was Mark's favorite part. His older Brother would always allow him to get whatever fast food he wanted.

When young Mark is left in the car alone, he begins being nosy. Like most Seven year olds do when left unattended. He sees that a gun is inside the glove compartment. He took apart the clip from the gun, but not realizing that there was still a bullet in the chamber he puts it to his head. He hears a voice tell him "Maybe you shouldn't." So he lowers the gun. But, he really wants to press the trigger at least once, so he aims and shoots.

The bullet quickly fires and ricochets throughout the car. The curtain lowers as the lights of my Moviemind returned to normal. Jesus than asks me "What do you remember about that moment?" I don't know how I completely forgot about that moment in my life. But, I did.

I remember waking up and hearing a loud ringing in my ears. There was seat stuffing floating all around the car. My vision was cloudy before slowly starting to clear. The ear ringing lasted about Two minutes before disipating. "You actually died in this world. My Father woke you up in the world that we just left."

As Jesus explains this to me, I notice the differences in where I am. At first glance it would appear that everything is as it has always been. I went on a lot of shopping sprees with God. Similar to the Walgreens visit I mentioned earlier. He had me buy two of everything. But, now in my apartment I don't have a duplicate of anything. I'd specifically bought Two, silver, metal trash cans, and now only have one. I can understand misplacing the headphones, or the slefiestick that I just bought. Even, perhaps AC stole them.

But, trash cans? Who stills or looses trash cans? It wasn't just inside the apartment either. My building is managed and mantained by more than Fifty people. Every single staff person had been changed. There was not a single person familiar at the leasing or, mantenance offices, or cashier working at any stores. I walked to see if my Barber was still the same, but his shop was closed. Closed on a Saturday? Black People, come on!

Only Two of my neighbors were the same. I didn't really ever socialize with any of them before. I remember Ms. Brenda on the Second Floor, because God had me randomly ask people on the street if they were believers or not. Ms. Brenda, was the only one to say no. Actually, she screeched "No!" With great conviction. And, I only remember Yessica from my floor, because she has the hot son, and. Last New Year Eve, I was tipping in, late, past her apartment. She had her door unlocked and was on one. Yup, "Turrnt up!"

After Yessica finished telling me that she would have my back. And, that I could count on her. And, that her door was always open, and that she got me. You know all the things we say when drunk dialing. Only, I'm looking right at her as she tells me that she was going to bring me a plate of food when it finished cooking. I waited up for hours, thinking that I would get some authentic, homemade Puerto Rican Cuisine. But, nope she's never been buy to drop a morssel of food. But, other than Yessica, and Ms. Brenda. I didn't recognize anyone.

Could it be that I really am some place different. Like I mentioned, I wasn't in much control of my body while all of this was happening. So, I just went along with everything only really believing in Jesus, and God. What's truly happening? Is this what it means to be a born again Christian? Where is everyone. Why did Jesus tell me to write down the things that reminded me of my Mother. Damn, I have to get back to Staten Island to see if she still is alive.

When I arrive, she is babysitting my Two nephews. Perfect since I had some gifts for them. But, I wanted to offer them the gifts as prizes won in their competing for them. But I arrived after Nine at night, and my Mom would not let me take them outside. What's with all the secrecy and departure of my family? My sisters didn't even tell me about their dangerous cross country trek. I had to learn about it from my dead nephew.

Now, here my Mom is telling me that. In the summer time. When not one of them has a pressing engagement they must be ready for in the Morning. I can not take my Nepews outside, in front of the house to be athletic and artistic? Participate in fun competitive sport? Do you really mean to tell me that if these children died this year, they will not have ever experienced a night, under the stars in the best outdoor temperatures?

I have a lot of questions to ask my Mom. Especially in regards to my childhood. She once told me that something happened to me when I was Seven years old during our residing in a home that I don't remember. She said that she would remind me one day, but it was too painful to talk about at the time we were discussing it. But, this clearly is not my Mom. This woman degrading me in front of my Nephews. So, I leave and return home.

Seven was the year that I was abducted by my Martian family, according to Jesus. That is the event that took place in effort to transition me between worlds when my Brother's bullet killed me. A Three year time span that I remember as a single day. A day where I as Mark rember it to be a day where I was playing in a wooded area, and I slice my toe. I was rushed to a hospital, strapped to a table and my toe was sowed back on, with nothing dulling the pain of it all.

For Omhmar, it must have been the trama that we do not remember, and that Mom doesn't want to talk about. Jesus reminds me that I must honor my Mother on Earth and my Father in heaven as he tries to convince me that my mother has transitioned. Mom as I knew her, was not my Mom.

As revelation and change become the prominent fixtures in my current dwelling. I feel as though I am being babysat. Each day by someone else, with specific knowledge to impart me with. On Sundays, both God and Jesus leave me, and I spend it with Jesus' Mom, Mary. Mary and Jesus are a lot alike. As if she is the female version of him. Which is half of the time bloody anoying.

Out of everyone she and Jesus get on my nerves the most. They offer me more guilt than anyone else does. However, if you are desperate for something, Mary is who you wanna turn to. Mary gives me whatever I want on Sundays, no matter what the purpose. Where as God only grants me things that are within our plan or discussed purposes.

He arrives on Monday each week, and blocks anyone else from visiting. Mondays are very spiritual for me as God and I have our alone time. It's usually the time where we evaluate the previous weeks. He always gloats as he shows me how I could have handled the situation better had I been channeling him, instead of myself. I become num to that part of his process after a while. There's never going to be a chance where you handle anything better than what God tells you he would have done.

At the end of the day, he's just talk. Because, when we triumphantly live to tell our stories. We tell stories of actions. This Monday, God has shown up to talk about who my husband will be. I've understood marriage always to be something other than you would. Never about love or romance. But, about status and estate planning.

God agrees. And, wants me to take some old ideas with me to my new world. Now that Mark's world will allow Gay men to marry one another. He would like us also to bennefit from some of its conceptual awards.

For most ritualistic ceremonies, which initialy is what Marriage is. There are spiritual components that justify natural and untaural bonds between the participants of the ceromonies. Do you really think that Elizabeth the First would have a passionate commitment to building Great Britain to become the Worlds leader, only to die and walk away from it? Absolutely not. Her commitment would live on through her transistion to the place where she would next exist. Existing in a place, exploring ways to advance her commitment in ways that we can not.

We typically will have movie night to end Mondays. This Monday's movie is Pride and Prejudice. I never choose the movies, and they all wind up being monumentally rewarding. This one stinks of a request. So now that you force me to get married, I have to let you choose? Son of a Bitch. Why is it that to whom much is given, much is required? Fuck no! Wait.

This actually could be a great idea. I don't really care who I marry. It's not about love and romance for us. I'm still going to fuck who ever I want, whenever. Not to mention, these people who watch my every move know what and who I need in my life. Remember the dream?

The usher with the purple scarf and the business contacts. Well, there giving me what ever I want. Finally all of the years of being a goodytwoshoe geek is paying off. Or maybe its reporations for killing myelf at seven. What ever the reason, I just hope that they include a McArthur Prize in it.

The week following God's departure was inccredible. I can't speak of most of it since secret society is just that. But, when God returns on Saturday, I am surprised that he has his son with him. "I thought you only got One week to show yourself to your chosen?" I ask Jesus. "I'm not sure why exactly. I can't find the twins, and they must travel with me." I know he's talking about Marcus and Kent, but I don't know what the hell he is talking about. God speaks, "I need to take you backwards with me for a while. We'll continue moving forward with my son afterwards."

Sunday when I wake, I am already out and walking. I don't know what the reason is, so I decide to go home. I am carrying a back pack with various things packed in it but I don't know what all exactly. But, what it didn't have in it were my keys. Is it possible that he brought me back to that day where I left my keys inside with the apartment door unlocked? I wait downstairs for a neighbor to enter and I would eneter with them.

As I enter, I am stopped by a security guard, "Do you live here?" I tell him yes, and he continues with questions such as, "Which apartment do you live in? How long have you lived here?" As, I am being badgered by this rent a cop, all the while the Latin people entering the building without a key are not being treated in the same way. I go off. I call him all kinds of names as I share my outrage for his racism.

He tells me, "Calm down or I will call the cops." "Call the fucking cops and I will wait right ouside for them to arrive. And, afterwards I will make sure you are fired." Yes, being chosen gives you a self worth that is often appaulling. Add to that, I am a Libra. If you know us, you know we don't go for lopsided justification.

I had a conversation with God once, in relation to America's escalation tension between Police and American Blacks. More specifically in Eric Garner's case. I couldn't understand why he wouldn't just follow the instruction given by the police. He couldn't quite tell me as best as he could show me. As the first car arrived, they walk right past me to go speak to the rent a cop. Shortly after that, Two more cars arrived, each holding Two officers. Finally, their sargent arrives, along with the neighborhood crowding around to witness.

"This woman says that you were taking pictures of the kids." One officer tells me. Before he could finish, "Do you live here?" Another, officer yells at my way. In a matter of seconds, I had Six police officers barking out accusations and questions. As they formed a simicircle around me, I didn't know who to look at or answer first. So I just hold my head down not saying anything. I don't know what Omhmar was up to before God switched me back to his body. But, clearly I had just fucked up big time.

Confused as hell, I use my inner voice to ask God, "What the hell am I supposed to do here?" I'm thinking, did Jesus remind me of the childhood suicide I had, so that I could remember how to take a bullet? Oh, No! Not God's chosen! Simply chosen to be another one gun down in the streets by the New York City Police Department?

Here we are now. I'm trying to get into my house, and the NYPD has sent out it's terror watch, training group to fight my terror.

God speaks, "Look up." And, I did. Witnessing the most gorgeous man that I had laid eyes on since moving back to the East Coast. Suddenly, like when I met Jeremiah and Old Stink Face in the hotel hallway. The background noise fades to sexy reggae riffs and rhythms.

I slowly look him over from toe to head. Stopping to savor the sweet aroma of my favorite parts. Thighs, chest, lips, teeth and eyes. He was medium brown complexion, with, as American southerners say. He had good hair. Baby waves, brushed forward in his dark fade. Oh yea, Officer Usher was all that, and a bag of chips!

"I'm not sure who to answer, nor when. So I've just been holding my head down waiting for the chance to answer." Is what I say to Officer Usher. While my inner voice was saying, "Yes I'll marry him." He begins the conversation that would end in me telling him my side of things. He leaves me briefly to speak to his Sargent. As he returns, the Sargent and other officers question the crowding neighbors. "I'm going to go up to his apartment and check it out." The Sargent tells my Officer Usher.

As the Sargent returns, so do some of my neigbors. Yessica tells the officers and Sargent that she knew me to live next door to her. They call the security office which doesn't have my lease on file. So God begins to explain, "Just calmly obey Officer Usher. He's going to take you down to the station for a bit, but you will be fine." I thought that I would only be going in to make a report, or something. When Officer Usher bent me over on the hood of his roof, felt me up, and bound my wrists tightly behind me, I learned quickly. It was not for dungeon play.

On my way to the station, I noticed that there are Movie trailors in the neighborhood. Odd, because I have never noticed anyone filming in my neighborhood before. The other night I walked past what appeared to be a rap party or some kind of gathering of the filming participants. I don't know. It may have just been the set dressing for the scene, provided that they were shooting outdoors.

Oddly enough, as I walked by the set. A man walks towards me as he crosses the street to where his car was parked. I could swear that the man was Tyler Perry. As he approached me, he kept repeating "He aint saying nothing." As he looked at me. I couldn't tell if he was talking to me or not. But, as he pulled off. I couldn't help but feel as though God was trying to bless me by positioning me right within direct access with one of my infamous heroes.

God may have been telling me to say something to him. I'm not quite sure. But, as I have obeyed God this time. And, now being escorted to the station in the caravan that has me cuffed and slumped.

I had never been arrested before. But, of course on the Cop shows, the cops always read the characters their miranda rights. However, I wasn't read a single right. So I assumed that I was not arrested, and was just going to the station in cooporation. I wondered if some how if I were chosen to star in a new Tyler Perry movie, that was using some kind of documentation where actors were not aware that their actions were being filmed.

There are many reasons why this idea would seem logical throughout the process. Most of those reasons, belong to my time spent with the council of the Illuminati. Which I have sworn secrecy to, and will not discuss them here. However, when I spent time with the Skull and Bones Council. The majority of our time together was spent with them teaching me mathmatical process I was unfamiliar with. Binary math to be more specific. I see a lot of the equations that they taught me, here etched in to the walls of the cold, depressing, prison cell that Officer Usher has just tossed me into.

I say wassup to my cell mates. One Black guy that appeared to be no more than Twenty Three years old. He was arrested on a routine, stop and frisk shake down of the American Black Man. For those of you unfmailiar with the policing practices of the NYPD. They have a system of randomly stopping Black men as they walk down the street, and making them hug a wall while spreading them.

For no reason at all the Black men are patted down and asked to show idnetification. Although they are violating our civil rights, right here in the world's mega city. This is completely legal, and encouraged by the Mayor's office.

This unfair practice will often land Black teenaged and young men in some of New York Government's finest, hospitable, accomodations. For simple infractions such as posessing a nickel bag of marijuana. Which was the case with Cellmate A.

Cellmate B, was attending the Puerto Rican Day Parade that celbrates his family's heritage. He was drinking a beer outside when a Cop put a haulting stop to his celebration. If ever there is a time that calls for enjoying a beer outside, this should be one of those times. Not for B's arresting officer, who took the parade as a vortex, incubating the exact atmosphere that would grant them some easy arrests. They could travel back and forth, all day to that single location picking petty criminal minorities up, by the van loads. And, they did.

I have been friends or on dates with White men, who move cartel weights of Crystal Meth and other drugs. Yet, they walk or drive by the Black men being violated and charged, as they continue to distribute the problems that the drug task forces are created to control.

Now, here the Three of us are. Sent to the Tower, for posessing a tool that allows you to escape harsh realities for minutes at a time, enjoying a beer on the parade route that promotes Puerto Rican pride, and tresspasing in the home that I am the lease holder for. That is what I am eventually charged for after spending Three days locked up.

I had no idea of what I was being arrested for even. I wasn't really clear of if I had even been arrested since I was never given my miranda rights. At some point in the God awful ordeal, I was transprted to some place where I had to be photgraphed, again. Checked in, again. When The lady checking me into the Second place I was transported to, asked me who my arresting officer was, my response "I don't have one." As I explained to her that I wasn't sure that I had been arrested, she assured me that I had been. And, that the Police having to read a person their miranda rights before arresting them was a myth, and they in fact do not have to.

I supposed my shy, uncomfortable, and nervous temperment may have rubbed the clerk the wrong way. She spoke to me as if she was appauled at how uninformed, and unfamiliar I was with the whole process. At this point they hadn't even told me what I was arrested for. Yet, the actions and harsh criticism from the clerk led me to believe that I was wasting her time, since. Surely, a Black man in the South Bronx, has had to have been arrested before, and I must be acting as if I am unaware of how this was supposed to go down.

Like, I was playing with here. Can you imagine? Me, a total stranger to this woman. Because of my skin color and age, could not have survived all of these years without a single arrest.

Maybe this place I am in now, might be central booking. Not sure, since they had Four of us shackled together by our feet, and brought over in a van with limited street views. So I did not see the outside of the building, and we were unloaded in a covered garage entrance as we entered the building that lacked definitive sinage.

The shackles were extremely painful, and would tighten with every step beyond the limited range of motion, that the shackles allowed for. By the time I arrived in the holding cells, my right ankle was swollen. The size of the swollen diformity of my lower right leg, seemed to be the hot topic being dicussed as we were handed off to the officers that will escort us to the next destination in this fiasco.

"Yo, look at your leg. You should go to the hospital." The dude arrested for the crumb of Marijuan, tells me. I look down at my leg to see what everyone was staring at. I can't see anything that would be unusual. I can't see that the leg is swollen. Nor, do I feel any pain. I'm not sure why everything seems normal to me, when everyone else is treating me like I just got shot in the leg.

I decide to take everyone"s suggestion. And, I request to be transported to the Emergency room. I didn't believe that anything was wrong. The witnesses to my leg injury must be actors, I assume. It seems that every setting that I am placed in, places me right at the center of attention. Now that the New York Government has requested an audience with me. I withdraw to my inner self, to play back some of the events I'd recently experienced.

Once we arrive at Lincoln Hospital. The absolute worst that the planet has to offer. I decide to request a Phsych Evaluation in addition to my X-rays. I must be going crazy if I can't see my own leg injury, which everyone els has described as really damaged. If my leg has suffered such damages, why don't I feel any pain at all?

A very beautiful, young African nurse preps my vitals and such. Once she is done, and I have to wait to be seen. I beg the officers to allow me to take a nap. At this point, we're coming to the end of day Two since detaining me. Every element of the arraingment process is constructed to keep you awake. I can tell you, that after Two days of being treated like a slave that is being starved, abusively injured, prevented from sleep. I would confess to murder, to get a bed right now.

The officers oblige my nap. But, they keep me in shackles and now they connect my hand cuffs to the bed railings so that I can't flee the country, or God forbid. I try to make a run for it, and their fat asses would actually have to get some exercise. Whatever the case, I am happy to finally be permitted to grab some sleep. While the officers get comfortable, nuzzeld in a couple of chairs with some magzines. Leaving the bed railings to do the policing for them. A grand way to waste tax payer resources.

About Six hours past. I have been X-rayed, and given blood to evaluate my Psycological state. I suppose that in New York, a Psyc Evaluation means, find out what dugs are in his system. Time to head back to prison. We pile back into the police van and head back to what I think might be Central Booking. As we make the short drive. All I could think of is my fucking Father in Heaven.

Why the fuck did he set me up like this? If I knew that I would be detained this long and under such inhumane conditions. I woud have never agreed to go into the precent. I wouldn't have stayed downstairs waiting for the cops.

Perhaps this could have played out a whole lot differently if the police had to pull me from my apartment, in order to arrest me. Silly of me to think that I could survive in the truth, that I too, have rights. My bad. The culture that is my life is typically absent of characteristics common to the American Black experience. So, sometimes, like in this instance, I forget that I am Black. Boy, what I would do right now, to be back in my Mozart, German body.

Once back in Booking, the clerks continue to process, and photograph me before placing me in one of many holding cells. There is a long row of cells, with Seven cells on each side. Each cell is packed with about Twenty guys. Of cource, I am placed in a cell that doesn't have a pay phone. Remember those? I haven't let my family know that I've been arrested, yet. When I arrived at Booking the first time, the clerk was nice enough to let me call Mom. But, when the voicemail answered, she didn't allow me to leave a message. I'm sure the NYPD has good reasons behind their processes. But, a great deal of it seems nonsensical to me.

I thought I would see my lil buddy that was in my last holding cell, back at the station. The one that was arrested for the few puffs of weed he had on him? Officer Usher was kind enough to allow me to lend him a shirt I had in my bag , since he was arrested wearing only socks and a thin pair of shorts. He'd been shivering the whole time. Part of the NYPD's effort to keep us awake, is to blast the AC to freezing temperatures.

For a lot of the detainees, tonight is a reuinion. "Club Clink Clink," if you will. They must have a system for seperating us. In the cell directly across from mine, there were only "Chulos." The type of young Mexican guys belonging to drug cartels and, waiting for the deportation bus.

Next to their cell was a mix of Spanish and Black women, currently detoxing. Probably arrested for prostitution. They looked old school, too. Remember back in the day when hookers would stroll The Boulevard wearing a napkin, and with a little more make-up, they could look like a Five? That's the look.

In my cell you could see that I wasn't the only Bueen. Hopefully this means that I'm placed in the safe cell. Most of us are dressed well. And, there are even a couple of White guys in it. There is a Black man dressed in all black with a very nice trench coat, and Purple tie. I am told that he is actually a Judge. There is one White guy that appears to be homeless sitting on a bench sleeping with his head resting on the metal bars. Can you believe that he did not wake up one time during the Fifteen hours I spent in Central Booking?

Finally, an officer comes to get me so that I can make my call in the Pay Phone Cell. "Hey Mom, thanks for accpeting the charges." She knows that it can't be good news from a collect call. "Mark, where are you at?" Mom asks. I tell her of today's events. "I'm going to find out where you are, and I'll be there when you see the Judge." Mom tells me before I hang up.

As I'm being escorted back to my holding cell, I noticed that items on the officer's desk look very similar to items I had in my back pack, when arrested. I'd bought a set of colored pens to identify what God and Jesus tells me in my notes, from my own notes. I'm sure the set could be a popular purchase. But, wouldn't it be unprofessional for the officers to be filling out government documents with colored ink?

I'm sure that the NYPD probably doesn't discriminate based on looks. But, the officers seem odd in appearance and behavior. One looked like Shanene from "Martin." Complete with long curly nails that were painted purple. Whenever food would be delivered, the officers would take milk and fruit for themselves and then hand us dry peanut butter sandwiches. To be so blaten. Then right at Eleven O'clock, imagine. My cell divided into Four corners. Each led by a thug clergyman who began to deliver a sermon to the followers. As I sat back in amazement. I wondered if this was normal, or if I were still on Candid Camera.

Many preachers speak of sordid pasts that included jail time. Could it be that I was witnessing the next round of teachers accepting their calling? I wondered if everyone in my cell had been chosen. There is one black guy that I keep running into during many of my challenges. He doesn't speak to me outside, and he is not speaking to me now. And, of course the White homeless man sleeps peacefully through it all.

Finally my name is called to go see my public defender. The holding cell they move me to next, has a minilla folder taped to the entrance of it. It reads, "PENS" in blue ink. I find out later that it stood for Pretty Entitled Niggers.

"You've been charged with tresspassing." Say's my lawyer. Can you believe it? Held three days. Injured, sleep deprived, starved to mal nutrition, and driven crazy. Taken from my own home and charged with tresppasing. God is about to get cursed out, Richard Pryor style!

The ordeal is almost over as I enter the courtroom. Excited to see my family here to support me, but as I look around, I only see their absence. The judge gives me a court day to come back and prove my innocense. My phone and Keys are still in lock up with all of my belongings. I find a pay phone to call Mom. "Hey baby." She answers. "Hey mom, I'm out now. Where were you guys?" I reply. "I'm on my way down to Maryland with your sister." There I have it.

Once again, when I need my Mom to show me that all the honor I give her is appreciated. She and my younger sister launches a plan to show me the exact opposite. I don't feel very chosen for much right now. Furious with God, I fall asleep depressed, cunfused, and angry.

The next day I take a break from God's process. The morning is quiet as I do some shopping, and just chill for a moment. There is something going on at the airport. I saw weird signs on 125th Street that read JFK closed and to use LaGuardia airport. They're were extra shuttles going to LaGuardia Airport as well. There were lines of men, women, and children boarding the shuttles with luggage in hand. When I get home I notice that the planes are flying very low. Then I question the frequency of the flights.

I'm a jet setter who's spent plenty of time on the tarmat to know. Planes leaving within Fives seconds of one another just doesn't happen. Flight after flight, every five seconds. Each one flying lower than the next. All day long, hundreds of flights flew directly above my building. As if they were all headed one direction, the hell out of New York.

God didn't tell me how long my training would last and when my missions would begin. And, Jesus is absolutely clueless, as to what the hell is going down. Jesus told me when he first arrived where the next terrorist attack would be in New York. I don't know if it has happened or not, but he tells me that now is the time to go warn the Mayor.

So I zip off to the train station, heading towards City Hall. As I get off the local train the express is coming into 125th to meet me. When the doors open a White guy, and A black guy, dressed almost identical to me. Definitely with the same color scheme, red and black exit the express train.

The three of us meet in the center of the platform, glairing at one another with confusion written all over our faces. They must be chosen too. As I enter the express train, they fall in behind me. Just before the doors close, I remember God telling me that none of his chosen should ever be in the same place at the same time, doing the same mission. So I barely make it off, as the doors slide close.

When I arrive back at my home station. It is poring like cats and dogs! The streets are flooded, and the water is pouring over the basins and grates of the subway station. I was only gone Twenty minutes at the most. It was completely sunny and clear when I left. Where did all this rain come from all of a sudden. Yet, it's not enough to ground these flights that are ominously speeding above my head as I run back to the apartment. Drenched! I'm like what the hell is going on?

Before any of Them spoke to me individualy. God prepared me very well to handle the conversations that I now participate in. Over the past few weeks, Jesus has taught me many revelating lessons in Theology. Definitive roles of his disciples, World History, and Secret Societies. He has even provided clarity for subjects that are purposely hidden from non-believers.

Hidden so well that, for now. I will have to take his word for it, as I believe. Because so much of his and their teachings are new, and require scientific comprehenshion beyond theory roots as we know them to exists. For that reason, among others. Validation is God's suggestion before action.

As we build this world to come, it is clear that I am living in my last days of the old world. I'd love to have my heroic time and history like Paul Revere. My time to sound the alarm of the danger to come in hopes that my warning could provoke the preventative measures that would keep a city with Ten Million people in it safe.

However, the logical part of me says that such acusations made without a creditable source, may land me back in Jail. I have no desire of going back there now or ever. I'm not sure that I am ready to be chosen yet. Here arrives my home girl, again just in time. The Spirit of Truth.

I begin asking her questions right away. "What's happening?" I ask. "An experiement," she replies. "Am I involved?" I ask. "Very much so." She replies. "Is it too late for me to stop?" I ask. "Yes."

As she answers, yes. I become uncomfortable again in my own home. I can feel things touching me. It feels as though I am with hundreds of things, spirits, Martians? I don't know what or who they are. I do know that something more than mysterious is happening.

Most days, I move about, without a care in the world. Positively, optimistic regarding all needs and desires. Hopefull in my Belief System, and Dreams. So when I do finally arrive at the day of worry. I often go to the extreme opposite spectrum arriving at some of the most negative, and discouraging assumptions.

I'm supposed to believe that I can't get it wrong, nor right. However, when I fail God, guilty and disappointing emotions consume my mind and body. Usually, I shut down, sleeping to escape the negativity that I can't shake. But, these days, sleep only takes me to spiritual interaction with whom do not exist in the physical. Almost as if I do not sleep.

I can't even begin to try sleeping in this scary, ominous environment. So I begin packing a bag, to hit the road. Not, sure where I'll head, yet. But the safest place to be at night during the Purple period is outside. Especially if I can see the moon. The moon and I can travel all night around the city, witnessing the miracles and scandals that happen only during Sweet in the Morning.

I guess noticing how frantic and confused I am as I gather some things for the night and tomorrow, Jesus intervenes. "There is nothing you should fear, but the Lord. Again, I tell you that as you become familiar with all that has unveiled. Your entire life was only for the purpose of this moment now. You can not get it wrong. You can not get it right. You have chosen, and been chosen. The storm is over now."

Dude fuck that! I keep doing everything that I am supposed to do, and yet it all feels very wrong. I have yet to see any of the rewards that were promised to me. I am working my ass off, day and night. Spending my own money to pay for experiments and lessons. That lead me where? Back to the starting point, that surpass concluding that this is bullshit.

I don't think that there is a place in New York City that would be able to ease my current state of mind. So, I head for Maryland to be with my family. "Before you go, I need to tell you that I am leaving and wont be returning for a while. While I am gone, study Einstein's Theory of Relativity, the Book of Mark, and for Six weeks study the number Six. Also, it is the Sumerians that Pastor Bernard was speking about. Not, Samaritans." Jesus commands as I head out of my apartment. "Deuces." I reply.

"Well, look who maid bail?" Diana says to me. Common in my family to make jokes before arguments begin. "Uncle Mark!" Exclaim, Diana's sons as they race across the living room, and up the forier to embrace me in the most loving, strong hug. For now I decide to let my hurt take a back seat to the true need of family. With us all living in sveral parts of America, it is so seldom that the immediate family is under one roof. Mom and her Brother. My Two Sisters and I. Their children and my Brother's oldest son. And, it just so happens to be Diana's youngest son's Birthday.

It's nice to see that my youngest nephew is devloping into his own character. Having them a year apart, my sister often treats them as if they are twins. Too often his older Brother has made all the opinions for him, and he just follows suit.

Now, at Nine. I can see that Denobenowitz, that's what I call him, has a cool personality all of his own. He's even gained the confidence to challenge the stupidity of JuJu. That's the oldest one. Ten years old. Denobenowitz has a few party guests. But, hardly enough to fill a room. And, I see that he's a little down, due to a great deal of his classmates not showing up. Meanwhile, the attendees are glued to their mobile gaming devices. I can't wait until the day that we're over it all, and get back to actually socializing with one another, at social events.

I can smell the Barbecue calling me to the back deck. And, it's time for smoke break. Tired from all of the traveling, I plop down into a comfy patio chair. Next to my older Sister. "I heard that you came down here looking for me." Older Sister say's to me. "Yea, Marcus told me not to let you get in the Car with Diana. But, you had already left for California." I answer.

"We didn't go to California. We were in Hawaii." Who does that? What would make a person lie to their own family about where and how they would vacation. My Two Sisters planned all year to visit Hawaii for their summer vacation, and kept a secret from everyone. Including their Children. I don't know about you, but anytime I board a plane. I want people to know it.

Furthermore, why did Marcus send me on a wild goose chase when their plan was never to drive across country. Perhaps his Mom was correct in saying that he's demonic. God's process is exausting and expensive all on it's own. Without, demons adding tasks and expenses. God, are you certain that I am succeeding?

Whatever the case, I made the right call by not going down to City Hall. Now these Bitches want to show me their pictures from Hawaii. The nerve! Do you know the hell I went through during the week they were missing. I've only told you the half of it, here. With my Sisters knowing most of it, pictures from all of their fun times should be the last thing they would shove down my throat.

Like a big pussy, I sit quietly, looking at each photo, that varies very little than its previous photo. Only speaking to say a couple of "How lovely's." Or, "Ooh that's nice." In the corner of my eye, I can see my Mom watching me with hesitation. She know's she just gave me another good example of how she always chooses my younger Sister over all of us. I get it. Diana's father was the only man my Mom truly loved. So, Diana will always be a little bit more special. Also, Diana is the only one with young children of the siblings, now that Marcus is gone.

I like to buy for my nieces and nephews, things from stores that their parents don't shop at. So before leaving I take Denobenowitz to the mall. For the first time in a nearly Two days, do I get to see that special smile on his face that every young boy should have on his Ninth birthday. You know the smile that says I can't wait to go back to school and show all my friends why my family loves me more than theirs. He felt special when I had him sized for European Shoe Sizes by a professional, and bought him Two pair of one of kind sneakers from trends that European kids enjoy. Than we shared a ride on the Merry Go Round. He had a blast, since JuJu wasn't around to tell us how lame the ride was.

I decided not to even mention anything to my Mom regarding my arrest. But, that wasn't enough to prevent her from insulting me further. When I asked her to make a spare key for me, so that I could go to her house and pick up some of my things. She, said to wait until she got back. She didn't feel comfrotable giving me a key. Nor, did she know when exactly, she'd be returning to New York. All I could say was, "Okay." As I gave her some money to susatin her remaining time in Maryland, grab my things, and push it back to New York. Back home to the place where I fear sleeping in. At night. Back to Mars.

Chapter 10: Like Minded and Worlds Apart

Bad enough that NYC is going through the apocolyps. But, the energy in my home still feels very wrong. I spend most of my days studying religon and Science. Per, Jesus' requests. I'm not sure if he has given me code, or something. But, I take his order to study the number Six, for Six weeks, to mean that I have to be introduced to you know who. I mean, he was onlly One, Six shy of Satan's area code.

Drudgin up the past to the origin of God's creation of evil to validate good, couldn't be the best energy to channel in my home. But, added to the odd feelings of extra terestrial life in my apartment. However, invisible. Still scared the living day lights out of me.

There is a Bodega directly across the street from my office window, that closes at Twelve, Forty Five, AM, every night. When they close up the shop, they pull down Three, iron gates. The noise from the lowering of their gates has become so irie, it acts like an Alarm to tell me that, if I have not made preparations to make my exit, I should be doing so now. As the Bodega lights are shut off, and the steet nearers the pitch blackness.

I have no clue why, but I feel as though God can not protect me when I am in the house. It could be that whole situation of when I had to go outside to send the Devil home. Whatever the case, I don't feel as though I am sheep, under God's glorious watch and protection unless I am outside.

While this period continues. I'm out of the house around this particular hour, everynight until sunrise. On weekends I may be lucky enough to stop in to a Party, or a hook up. But, during the week. Trying to be the Christ-like obedient servant that I am. I stay far away from trouble. Surprisingly, there is still never a dull moment. Even with me fleeing fearfully, with many nights of me not making plans to occupy my exaustive process that burns the midnight oil. We shop mostly.

It has been months since I have been begging Mom to go with me to pick out a car while I am still holding on to her down payment. I notice that in the time that she has avoided even looking for a car, I could have invested the money into Griffin Cleft, and turned it into a profit equal to, if not more than the principle that I've placed aside for her.

But, now each night I pinch off of it, as God and I shop. Everything that I do comes along with some kind of lesson. Frustrating most of the time to have someone in you ear all the time with an explanation of why to make a purchase, or add an ingredient to a dish I am preparing. Yet, I have to admit. The food tastes better, and God has hipped me to the action of making purchases simply for having surplus product. Product that you can then return back to the retailor for the full refund when the coins get tight.

Such a simple logic, that I would never have thought of doing myself.

Mon, being the single provider that she was, as we grew up. Never had enough money to buy surplus of anything. And, perhaps if a father or Two had stuck around to raise me as a young man. He may have hipped me to survival tips of how to live the best quality of life possible, while being short on cash do to building an empire.

As God introduced himself to me as my Father in Heaven, we have really gelled into a Father/Son relationship. That seems to have been old hat for us. Since he's blessed me with his process, I often feel as though my actual Father died and went to heaven. And, now communicates to me from the otherside, where we now together continue the work of his family business that I've inherited, operate, and continue the foundations and missions he began. For instance.

When I disbanded the Los Angeles Modern and Ballet Company. My initial intent was to move the company to Atlanta. Where I speculated I would gain better support from the large population of African Americans that would support Black people trying to produce and perform classical artistry.

But, as that attempt failed poorly. I had no intention of ever creating a Ballet company again. However, it has become his will in my life to create yet, another Ballet company. I was clear with him that I woul not take a prominent leadership role in the company. However, I would set the company up, while adding it to the umbrella of Griffin Cleft.

After all, as he points out, "Your mission of sustaining opportunities for Black Ballerinas to obtain gainful employment in the Classical arts. Is not a mission that has been remedied. Black young ladies are still faced with the same challenges of performing the craft, they spent years of hard work, and a fortune developing. Someone must continue the conversation and inact efforts that resolve sollution. Or else, you will wake up to a world one day, where young Black children would not ever be seen in a Ballet class. Simply because it were a waste of their time and hopes."

There is an abundance of luxurious existence, that exists seemingly effortlessly. Like magicians, and the worlds greatest masters of illusion. There are Human Beings that have accepted their spiritual callings, accept and agree with their purpose for living, and work tirelessly to make sure that a good quality of life will sustain for all that dare to partake of it. Have you heard of George Balanchine?

I know that Ms. Wendy doesn't really include stories with substance in her lists of "Hot Topics." Before the 70's, fame was readily available to anyone who dare to dream big. Although, these days, if you don't have a reality show crew following, and recording your every move for the world to pick apart your entire life.

Fame seems to be far reaching. But, in the 30's and 40's, classical artists, could not only exist in their places of true positive creation. But, they could also grab as much money, fame and power as Bethany Frankel. But, with out negative selling out, that is required to achive such a feat today.

The newspapers were obsessed with them in the same way they obsess over Taylor Swit today. The artists of the stage were the absolute romantic, heart breakers. Whom often entered multiple marriages, scandals, left legacies that would leave you gagging. I mean there are infamous stories of Balanchine's fierce terror, such as firing a First-rate, Prima Balerina, only minutes before her curtain is to rise. Not bad enough yet? Eventhough, he didn't have good reason for it, other than the Prima balerina filling for divorce from ther tremotous marriage?

Okay, still not bad enough? How about, he didn't even tell her? She was there, warm, face painted for the Gods, and ready to kill it on the stage. As she went to dress in the tu tu that she was to perform in that evening. Here dresser entered her dressing room and removed the tu tu from the costume rack on which it hung. And, only uttered the words, "You wont be needing this tonight."

Aside from the legacy that makes you gag. And before I froget, Mr. B, as he is affectionately called from those who love him. Mr. B defected from Russia to America in a time where such actions were considered treasonist and the truest of discracing his homeland. But, who could blame him. He was the hot shit, and so was America. How could he refuse such an invitation from Mr. Lincoln Kirsten.

As American Capitalism was starting to gain its stride, Americans had more money than they knew what to do with. The demand to be entertained catapulted the careers of international snesations, and out of the box thinkers like Balanchine. Why was, he so out of the box? It's just ballet, aint it? I know that's what you are thinking. Ballet has been around for centuries before his abuse arrived in the states.

George Balanchine founded the New York City Ballet. The modest begninigs that began in a High School Auditorium, would one day become the world renowned and celebrated official company of New York City, and NYC's Lincoln Center. The mainstage of Mercedes Benz's Fasion Week. Before Naomi stomped the catwalks there, the New York City Ballet Company put Lincoln Center on the map. Audineces from all over the worl flocked to the city to see what his premiers on each season would be like. One of the greatest chorégraphers to ever live, and made a seemingly small contribution that would change Theatrical productions for ever.

At the time that he bagn the company. Corps de ballets, which is the girls who just stand in the back. They would do just that. Take a couple of steps and, then strike a pose. Take a couple more steps, and then strike another one. The reason was due to such limited space when you pack Eighty girls onto one stage with big fluffy tu tus that aren't suppose to touch.

Unbeleivable that no other choreographer challenged convetional limitations by asking the question of how can we make the Ballerinas standing in the back be more interesting when the Prima Ballerina needs a rest.

His sollution is the fundamental base of what is now know as the Aerican School of Ballet. The French began Ballet, and other nations have adopted slight changes and approaches to performing theis graciously beautiful artform. In Mr. B's American school, he made his dancers stretch to become extremly turned out. Turned out is that thing that makes dancers walk like ducks.

What he was able to do with this new ability, was create a ballet technique that made the dancers perform in a small imaginary box, by implemting overcrossing of their feet. Allowing the dancers to move forward in backwards in a single Twelve inch by Twelve inch square, that gave them their own personal real estate on the stage.

Thus allowing for Sixty, plus dancers to performing exhillerating choreography all at once. Never seen before was a stage full of dancers twirlling for their lives, instead of taking a couple of steps, and then striking a pose. That is where America, and Russia's reluctant gift of George Balanchine, birth the Chorus line.

The rockets, majorette drill teams, parades, and so much more. Now exists in a much more exciting way, than how they appeared before 1940. His technique was applied to Vaudville, Opera, Broadway, Hollywood Films, and NFL Halftime Shows. Ladies and gentlemen, you can thank him, and the hard work of the dancers from New York City Ballet for your beloved Kicklines, and such.

In the 40's, the question was more, how can we make classical performances more interesting to the general public that were not afficionados of the arts. And less, of how do we make it more inclusive of minorities, and the Ninty Nine percenters of the world that lack access to the expensive and exclusive training required to succeed at professional levels.

That is more of the problem today, now that Ballet and Opera have bitten into some of the inner city youth that has ventured into a performance on a class field trip. And, my question, more specifically. Is, how the fuck are we going to pay for this shit. Producing Ballet and Opera is very expensive and requires as much effort to fundraising as a poiltician, when running a presidential campaign. In this new world where Government support is strinking, and the children of greatly endowed philanthropic institutions would rather feed the kids of Africa, than to continue the support of the classical arts.

Support that their ancestors new would be greatly needed. The hardest part of being the Artistic Director of Lambco, was that it would be hard to look a funder in the face. A funder that had a desk stacked with worthy proposals of great needs. Such as, HIV research, Cancer research, disaster relief for gulf nations such as Hati, as well as they forgotten Africa. And, then ask them to give me One Hundred Thousand dollars so that I can produce a single performance that would only be witnessed by the audience in attendace to the Three Thousand Seat Theater or opera house.

But, shamefully. Art's organisations are up against all those tear jerking, worthy causes. Autism, and mental illnes support, etc. Categorically unusual bed fellows, yet we are all competing for the same dollars, fueld by a small source of financial resources. But, God is right. Some how some way, I have to do my part in sustaing opportunities for the Black Ballerinas of today and and to come. So, yea. There's that.

We haven't had a conversation yet about my obligation to belong to a church home. My Family's CCC, led by Pastor Bernard is about a Ninty minute commute from my new digs. Appauling that in a city with Ten Million people living in it. They've labeled Sunday's as weekend days, requiring less demand for public transportation. Thus providing less frequent, local traveling trains that are inconveniently inclusive of track mantanence, that render certain stations unaccesible, all together.

I lack the interest to put myself through it every Sunday to go to a church that does a great job of filming and autibly recording Pastor Bernard's teachings. Even before I made the choice to return to Christianity, I had a strong interest in attending a service at Time Square Church. Sitting here and doing a little research. It doesn't seem likely that this would be my kind of leadership.

The Pastor is a little bit more mid western than I like my Pastors to be. After deeper exploration of their web site, I come across some select services uploaded from Times Square Church's, assitant and Youth Pastor. As the Law attraction would have it, of course I would click on the one sermon that spoke directly to me from said Pastor.

Already intrigued, because he was a young Black man with loads of swagger. Future Obama, kind of shit. As he began his teaching, he focused on a story from Job. Where the spirit of truth had shown up to deliver a message from God. Almost word for word, depicting a happening quite comparable to my own story that was currently taking ceter stage of my life.

In his story, of course the Spirit of Truth is a Man. I am not sure if since we're developing a new world, perhaps God may have made suitable casting changes that would work specifically within my own acceptance. Yet, there is the idea that the Church, back in those days would know damn well that the Sirit of Truth has always been a woman. Yet, they could not give public acknowledgement to God choosing a woman to do anything other than cook, clean, and bare children.

You would think that this moment would culminate the truest validation that this is really happening to me. At first introduction, I was so happy to be a believer again that I would have believed anything.

Belief was the farthest thing from my radar. The awful feeling provoked from being in my apartment at night, gave me all kinds of crazy speculations as to what and why this was happening. Now, in 2015 God desperately needs my help and is going to all the trouble to convince me that this is real.

Some of the shit is so crazy that I'm left wondering. Well if you can make squirrels run up some stairs, and then turn in to pigeons that take flight. Why can't you just morph into a secret service type dude, cleary present me with a mission, contract, non dislosure agreement, and a suitcase of unmarked bills?

When I ask God how the super natural wonders are infused into the physical world. He doesn't understand how someone as loyal, and stands in the truth that is belief. Could doubt that God has the power to create a world that only I can witness. As it unfolds right in plain view of everyone else to see. Yet, they don't.

Like any detective worth their weight in salt. I am constantly attempting to piece together clues that will lead me to who is really responisble for what is happening to me. I can agree that a great deal has to do with Spiritual Warfare. But, other moments how dumbfounding!

Each night I fall asllep debating throughout my mind of who could be behind the days events and displays of devine planning. Some days it's the Catholic church. So it didn't seem all that unlogical that they could have known that I would be moving to East 141st Street. And, before I arrived, they wired the entire building, and apartment to feel and hear certain things.

Science and technology continues to come a long way. I've also thought, more than once I am afraid. That, since my building has an uncanning number of deaths, my apartment could be housing very special spirits who are oddly capable. And, posessed a deep fondness for me, and concocted this stupid process to try to help me return to the glory of who I use to be. As, New York turned me jaded, as it is known to do when sucking the happiness right out of your body.

So this night, I fall asleep strangely doubtful that God is doing this. My blame now shifts to the U. S. Government some how. After hearing the story about the original Spirit of Truth, I wondered if they're just banking on my stupidity of the Bible. And, if anyone else who was very familiar with the visit she once gave to Job, wouldn't I have fallen for it?

But, I didn't just fall for her words and enegry. When she validates the truth, she sends these orgasmic, goose bumps, tinggling though my body. The feeling is so deliberately provoked that, not any of the others, able to offer me the same validating feelings in an identical way.

But, the conspiracist in me wants to know. Is it possible that there is some kind of weather machine that can be piped through my vents like a nerve gas, that would then make me feel and hear things manipulated by Scientist working for Uncle Sam. Or, the Pentagon. I've heard that the U. S. Government is infact involved to an extent. However, if this is entierly happening due to Government control, they would be spending a great deal of money on intellegence, staff, survelence, and more.

Of course they got it like that. But, why would they spend such a budget on a no body like me? Then of course the answer to myself would be that, once they find out that I am chosing by God. A special interst of me grows, not to mention. They'd think that by tracking me, they would actually be keeping tabs on my Father in Heaven.

None the less, I don't just happen to stumble onto my feelings. In fact, Abraham tells us that our feelings are the antena's, which indicate what is right for us. Since, I have wanted to visit the Times Square Church after my return to the East Coast. In fact, I can remember on visits, many years ago, wanting take a peak in. I suspect that it's quite the sanctuary, based off of namesake, and location alone.

Church's with artist ministries are my favorite. I hate to attend a church service where the praise and worship lacks any kind of professional standard, or techniques. So what the heck? Let's throw caution to the wind and venture to see what this old, White Mid Western preacher has to teach us on Sunday.

Maybe he'll be preaching from the Book of Mark. In addition to exploring Satan's origin. I've been studying who Mark originally was designed to become. Most odd to me, is that Mark's full name was John Mark. I wondered if Mark was truly the combination of both John and Mark, true disciple of Jesus Christ. A similar process that God claims will be happening to me soon.

I'm not sure of the trick that he has up his sleeves. But, he's prepared me to know that an aditional spirit would be making claim to the left side of my body, as its new residence.

My Griffin Cleft new hires, whom are both Black. Will often not show up for work, and offer me the same excuse. They were detained, citated, or bruttaly locked away for some dumb shit. And, for days at a time, without a single call to explain their absence. As they described their unfortunate run ins with the NYPD. I totally understand and accept their reasoning. Since I now, am very familiar with what we brothers face, north of Central Park.

Granted. There is evil that exists. There are harsh truths that may affect me, just because of the color of my skin. And, no. Most days don't come in with a bed of roses for a lot of people in this country that looks like my family, and I. I know that I can sometimes turn people off, always having an exuberant sunny disposition. But, in my quest to adopt beliefs that are believable One Hundred percent of time, seemed to never end. But, now I have discovered Abraham's philosophy of the Law of Atraction that works in my favor almost all of the time. Clearly my advanced grasping and implemntation of the principles taught by Esther Hicks greatly is responsible for my attraction of God, Jesus, and his accompanying spirits.

None of this wonderful time of constant creation and evolution could have been possible, if I we all were not matched perfectly in our vibration.

When I lived on West 137th Street. Abraham gave me a stunning visual of what was to come.

Cut to a year later in my new dwelling, God unveils what he has done. The reasons for it. And, has chosen me to play an instrumental role in the creation of our new world. One that shows real promise for finally obtaining world peace. I'm not sure if Abraham had any idea of what was to come at the time. It is sometimes painfully hard to figure out which sector of Them, believes that there is a God.

I'm finding evidnece that if you are a scientist in this world, you are a scientist on the other side of it. Meanwhile, I can't quite figure out exactly how, when or what was actually done for me. Yet, today I wake. No longer on Mars. God tells me that we have arrived on Father Earth. A smaller version of earth uninclusive of all the nations and places that make up Mother Earth.

But, with basic duplication of Mother Earth's largest cities. Most of us wouldn't speculate, that all the destinations listed on a globe didn't truly exist. I mean. If I won a sweepstake to spend Seven days, and Seven nights, all expenses paid to visit Hati. I would stay home, rather to watch T.V.

The purpose of this world we're in now is to validate what is to come to Mother Earth. A check point to ease some of the frustration, ideas, or events that would certainly be deemed as failure, or hardships if they made it to the Motherland. I guess you could say the Men looking out for the Women.

Yet, when God tells me that Ladies always go first, yet Father Earth is the place to validate what happens on Mother Earth. It confuses me to the point of not understanding which world comes first. But, since I know that I am going to be living in the newest of our Universal places. Who gives a rat's ass? Let's hurry up and complete the task mission, and raise up the hell out of here.

God speaks, "Tomorrow this world will be holding the American Presidential Election. You will help us to decide who will lead the Free World, next." After he's done with his introductory briefing me on Father

Earth. I'm led on an quick study education on American politics, and what political parties stand for or represent. Here is a point where I shock God with my choice.

He, and everyone was certain that I would continue to vote with my democratic upbringing that has dictated my choice any other time. I mostly, wait until the last week of the race to see who is leading in the polls. Which ever the clear winner is predicted to be, is the candidate that I normally vote for. Don't wanna get to the water cooler the next day to admit that I voted for the looser.

Here for the very first time God challenges me to take some issues to heart, and vote on the candidate that is most suited for my own political desires. "Where would you like to see your generation go? Who is most ideal to lead America in Mark's World?" God asks, me. I did him one better, which may have been the shock that turned my events toward a direction that he'd not intended for me.

I decided to change my party affiliation to become a member of the Republican Party. With Black lives mattering, and all of the evidence that points to harsh systemic design for minorities, should have led me right to becoming Hilary's bitch. Chelsea Clinton and I were at Washington Ballet at the same time, and I have danced for her parents many times.

But, as I gained deeper knowledge of the things that Republicans stood for. The fondness I hold for President Clinton wasn't enough to side with the Democrats, this time. A big appeal to voting Republican, is small Government. My biggest issue personally, right now is my series of unfortunate events with the NYPD.

The NYPD can't do anything to help me get an order of protection against my gay basher. Yet, have the resources for Six officers, and a Sargent to unlawfully arrest me, violate my civil rights and take me from my home to be abused in their antiquated, broke down processing and survelence.

I forgot to tell you that while I was under the clink. Those motherfuckers cloned my cell phones. And, for such inhumane treatment, my hard earned, and much needed taxes go to pay their lavish salaries. And their union makes it virtually impossible for any of them to loose their jobs.

Well! With a Republican regieme, the country will have no choice. Their payroll budgets will be slashed and, our taxes wouldn't afford them opportunities to bankroll the opressive enforcement of racial divide, misconduct, and abuse. Meanwhile the Democrats wont do much, but pay for lavish appreciation banquets that honor their commitments, of ensuring that American Minorities will always live in a police state.

Secondly, I feel that the system that administers entitlement grants and programs needs a major overhaul. We really could use a design that efficiently guides recipients to regain their independence. While the dilapidated, Democratic system is designed to keep its recipients relyant upon such a poor quality of life.

Tomorrow's validating election that would predict next year's election of the Forty Fifth President of USA, on Mother Earth includes all of the candidates in Next year's Primary Election. However, it's a General Election here. First let me say that, we celebrate on July Fourth, America's independency from Monarchy rule.

Therefor, the appauling idea that our nation could be controled by only Two families for more than Two decades scares me enough to cancel out both Jeb Bush and Hilary Clinton. But, I did enjoy being part of making history with Obama. And, could join the American's who would like to once again make history, by electing America's first Female President.

After watching a debate of all the Republican Candidates, Carly Fiorina was the winner in my mind. How fortunate that my initial gut feeling for the only possible female Candidate, was also the clear choice overall. Sadly, once I familiarized myself with her background. I learned that most Republicans weren't going to support a Woman who ran HP into the ground, for president. Maybe they'd forgive a White Man. But, certainly not a Woman whom they've had eyes on for many years and hoped that she would fail.

So Donald Trump it is!

I couldn't fall asleep last night, because of all the excitement related to learning new truths. I suspect that I am required to fall asleep in order to awake in the propper relm that I am supposed to exist in. Since, I don't have any more demands to complete on Father Earth, I am stressing to fall asleep. By now, I have given up trying to figure it out. Whatever it is.

If any of this is not making since to you. It makes even less since to me. I am pleased by it all, since I feel the love of God in varying excalation now. Sleep is not happening, so I get dressed and decide to attend the early morning service at Times Square Church.

Odd to be excited to go to church again. But, here I am about to burst with eagerness to visit my choice for a home church, for the first time. On the way downtown, I took the opportunity to shoot some promo segments for Mad Science. I was dressed for the Nines, so I might as make this day even more purposeful.

Walking through New York's, Broadway Theater District, just after sunrise gives me a nice feel of the city. I usually hate that part of town during show times. It's full of crowded streets of tourist and noise. I am greeted by a tall, sexy, White Man that appeared to be late 30's, when I arrive at the entrance to the church. That's always a great way to start an impression for me.

Henry Purcell begins playing in my mind, as I enter the sanctuary. Remember. Remember not. Lord, our offences. Is what the chorus of angelic voices tweedled in my ear while I march slowly down a long, red carpeted theater isle. I had a feeling that it would be designed to match similar designs of its neighboring Broadway houses. But, I had no idea that it would be so immaculately, omnipotently, ornate.

Mountains of luxurious plush seating that seemed to never end. A very regal color palate softly lightened, more and more as the walls met, a ceiling mural that seemed to be commissioned by Louis XIV. All I could think of as I took in all of the opulence was that Mom would be so proud. I have finally found my home. Imagine someone that would die if he couldn't praise God, or be on The Stage anymore. Now close your eyes and imagine what his ideal church would look like if he had total control of what it would look like. That's exactly right.

Only now, although I'm happy to see that the congregation is not waiting for church to begin before they pay, or praise God. I am disappointed that this service is so sparsley attended. WTF? I can't be a part of a church built for Two Thousand people to attend service, with only about Fifty other people with me. The Pastor can be as energetic as Richard Simmons, and it stil wouldn't be enough to deliver exhilleration.

I go oustide to check the schedule on the Entrance Manifesto. When I find out that the service that I thought was schedule for Nine O'clock, is now scheduled for Ten O'clock, I could kick something! Because, I was already an hour early for what now will be a Two hour wait for service to begin. After checking the web site to see if they needed an update, I confirm that the schedule was in fact correct on the Manifesto.

I was certain that the web-site listed the Services as Eight, and Eleven O'clock. Now they've not only changed the times , but have added Two more services. I want to be a good Christian, but asking a young man to wait Two hours for a church service to start after not having any sleep in more than Twenty Four hours is just too much to ask. Deuces, JC.

On my ride back uptown I begin to wonder if I had transissioned worlds again. I know that the Holy Father is capable of a great many things. But, I try hard, shifting through a backwards scan of my morning to see if I could fnd a moment that made since for me to leave one world and enter another one wide awake. "You don't remember do you?" They ask me. "Remember what?" I ask. "You fell asleep on your way downtown for only a few seconds. That was all the time that I needed." God Explains to me.

"You son of a bitch! You knew that I would arrive too early to stay for the service. Was it that important for you to make the switch?" I say loud enough for the other passengers to give me the, is he crazy, look. "We couldn't risk you befriending someone who doesn't exist in your world." They say. "Besides, you're normally late for everything. Who knew that you would actually be on time for something?" Sarcasm, added by God.

When he's right, he's right. I wonder how much of God's insipration is intentional. Often he'll come down my to my level to throw shady sarcasm back at me. Sometimes as he does, I'll pick up on secret answers to questions he wouldn't answer when I asked him outright. Like a recent ask I offered him about how much of the future he knows about.

Most Christians are in the belief that he knows any, and all to come. However, I've seldomly thought that, if we are given free will on Earth. How would he know what we do with it. If he just told me that he didn't think I would be early for church, and gave him a surprise. He's confirmed that he doesn't know the future anymore than I do.

I've been counting down the days until I would get to see my beautiful Officer Usher again, and it's finally arrived. Today is my day to see the judge regarding that awful ordeal where the NYPD kidnapped me for that weekend. Then charged me with trespassing in my onw home. Cleverly orchestrated by my own lease management and residence security professionals. You remember. Well, time to see if the judge is gonna make the charges stick. Fucking New York!

At least now I can feel Officer Usher out. I do believe that he was into me when he arrested me. Hopefully, he's gay and not on the DL, wanting only to fuck me on the weeks his wife wont give him none. God and I had a conversation where I told him that, esteticlly. Officer Usher would make a good match for my new husband.

He wants desperately to find the man I will marry, for me. And, get back to it's religious origin of creating a relationship where the bond is formed for the couple first. Then, over time the union unveils to them. Timely introducing charcter, ability, challenge, success, rebirths, while the love and commitment grows only stronger throughout their lives.

Since I've agreed to allow him this opportunity. Let's face it. I was among the Gay's protesting to save the sanctaty of marriage by keeping it between the opposite sexes. I hate being told the what, but left in the dark as to when the "what," will manifest. As I run frantically to be on time. I know what you are probably thinking, I caught the wrong bus.

Getting adjusted to traveling through worlds that appear to be identical requires careful travel planning. I arrive at the security check point, with a line out the door. I suppose, that discrediting poor folks would be an insanely, busy establishment.

I look around at all the officers, comparing them to my Officer Usher. Manogomy is not something that interests me, and I don't know yet if it interests God. I would imagine if that he were to provide me with perfection, what more do I need. Nope. I've decided that none of the others, are as sexy as my Officer Usher. That's a good sign, since they were all in uniforms that made them quite appealing to me. I think I'm sure that if any of them tried to get at me, I would still go home to Officer Usher.

I go downstairs to find the court room with my judge and attorney in it. There is a crowd of young black kids in front of my court room when I find it. I know the look of a dance troupe when I see one. There not in costume, but each one of them has on some kind of loose fitting style of clothing, as they congregate standing in First and Second positions. You know the way dancers must stand in public, to let everyone know that they are better than them, because they have studied the classical arts?

Then, someone appearing to be a dance coach, or teacher sticks her head out of the courtroom next to mine, and summons them to all come in. As, they file into their room, my attorney enters the hall where they once stood, from my courtroom.

My attorney, a dead ringer for Ms. Jennifer Aniston, quickly explains how the prcess will unfold today. I try to focus on what she is saying, but my nerves escalating to a quiet rage. All I can think of is, he better not be doing this to me. I think God has somehow orchestrated my wedding to Officer Usher. I told him that I wanted an event wedding worthy of being on that Flash Wedding show.

I was arrested on some bullshit, after God sent me backwards, just so that I would experience the bullshit arrest. When a copy of my lease can be mailed to the court to prove my innocense, I am summoned to appear live. And, I show up to see the cast of The Color Purple, warming up in the hall outside my courtroom.

Just putting this all together, and wondering if the courtroom next to mine was really just a holding area to unveil the surprise once I am in the courtroom. "Mark Usher." One of the court officers yell out. "Okay let's go." Attorney Aniston, tells me. I get up in follow her in to see the judge. But, not before asking why the officer called me Mark Usher. She claims not to know. I've already told God, that my husband will have to take my name. So they're already fucking up from the jump.

Expecting to see my family as I turn the corner, and my honey. I don't recognize a single person in the room. Yet, as I wait for my turn to see Your Honor, Three more officers have joined in on calling me Mark Usher. Which is clearly a combination of me and my arresting officer's last name.

Would you believe that the trial completed just Ten minutes after I got in there with no sight of Officer Usher. I thought at least he was required to show up to offer his testimony. The addition of all the oddities made me certain that he and I would be together today. Again, finally.

Attorney Aniston tells me that I might have to come back in a couple of weeks once she decides on my case. There it is folks. Me sitting in a long, dim hallway alone, which only Twenty minutes was filled with what seem to be my wedding party. All vanished into thin air like abra-cadaba.

Once again I was left feeling like it was all bullshit, and that I may be making it all up in my head and need to seek out some mental therapy. He did tell me that I would need therapy when it was all said and done. But, we'll have time to talk about that later.

My electronics have calmed down a bit, by the way. I had a conversation with God to ask him to take Marcus home. I could feel that he was too excited to be a part of what we were developing. It seemed as though he did nothing but try to get my attention. But, then when I would give it to him, he would only start a new way of trying to get my attention.

He offered no more than providing me with evidence that communication between two worlds was possible. It became a redundant finding not worthy of the constant pain in my ass he became. AC told me in a dream that Marcus was coming back as a little girl. I suspect that, in the future. I will be meeting a little girl that develops a strange attachment to me.

I found out that, the mother that visited me in that dream, actually was the

The Purple Book

Mom of this little cutie that I call my younger brother. I call him my younger brother, because his grind reminds me of how I was on my grind when I was his age. Rare to see in New York, young Black men know exactly what they want in life, and then charge full speed ahead.

I created my first private company when I was only Nineteen years old. And, been knocking out new ventures every since. I got the opportunity to meet Micah's mother only Once, before she passed. But, I figure that was enough for her. She had been living miserably, while taking care of Micah and his Four Sisters. Perhaps she knew that Micah would be in good hands if she went on and got some rest.

Sadly, most gay men live ther entire lives single, and lacking the since of security that gives parents the comfort of know that their babies wont be left all alone in this world when they make their transistions to their next world. I bet if you're reading this and have a Gay son. You've never even met any of his friends. I still remember the big, infectous smile that Micah's Mother greeted me with when I met her.

I couldn't tell if she was trying hold back laughter as I sontered into her bedroom with my fabulous ways. I love to serve it up extra, when I get the parent introduction. It's not hard to be impressive in New York, since most of the Black guys are uneducated, boring clones that try to act like L. L. Cool J. But without his swagger, money, or confidence.

I vaguely remember the dream where Micah's Mother popped in for a chat. But, I'm sure it was among the same line of asking that all mothers shovel at me. So that I can then go forth, being the messenger to their Sons from the otherside. I've known Micah for about Three years. And for those years I have been trying to get him to work for me. He is certain that his passion is Finance. Meanwhile, he dosen't posses a single trait of someone that would be succeeding in the Finance industry. Not even an accountant.

I've always taken this to mean that there is a small window that could afford me the opportunity to attract him to the enetertainment industry. Half of entertainment jobs have to deal with finace anyway. I've even told him of how I could tailor his job duties to include more corporate finance tasks, and less artistic entertaiment tasks.

He has always told me of how much he has admired my business, and that he sees me achieving all of my career goals. But, yet has declined to accept any of my offerings of employment. That is until now.

Micah's mother just so happened to make her transition while Micah was in between jobs. Bewteen you and I, Micah most often resides in between them. His Mom took care of him by leaving the apartment. It's a very big, spacious, Three Bedroom that's located in prime Bronx real estate, with great professional management. You would be hard pressed to find another Black dude in his early Twenties that would have such a jewel.

The ink hasn't even dried yet on Micah's Mother's death certificate, and already his older sister is trying to take it from him. Micah came over one night recently, just to hang out. I can hardly pretend to care, as he tells me how hard his struggle has now become with out Mom or a Job. Don't want to sound insensitive. Whenever peple want to waste my time with grand ventting, I sit quitely wishing that we can skip forward to part where we can debate sollutions to the problems.

I've got the perfect sollution that can put Micah to work as we speak. Griffin Cleft is hirirng. After years of courting, just like that I got him to accept my offer of employment. I hope that gives his Mom some resolve. I know that leaving Micah the apartment was an act of deliberate purpose. I hope that he is able to keep the apartment.

Like most children of single mothers, Micah was raised in the church. From our coversations I can tell he is a believer. Black single mother's are excellent at training up a child in the way that they should preech. But, raising them to be capable, prepared, decisive, consistant, strong men, is where their children get shorted.

To think of starting your life without your single mother being around anymore is tragic. Especially if she didn't properly prepare you take care of yourself. But, there was a time in America where Eighteen meant time to find a husband for a woman. And, time to get higher education that allows for good paying job, for men.

So sadly, Micah joins the overflowing club of adult Black Men, who are completely clueless. The Black Men in America who were raised by only their Mothers when their deadbeat dads, fall into the opression trap laid for them. Jeremiah and I both chose a profession where our careers bagan when we were teenagers.

So, when he reached his adulthood, he could not only take care of himself. He paid all the bills for both he and his mom. Unfortunately, Micaha chose a career that requires a great deal of education, money and connection to succeed in. All he has to help him reach his goal is a second rate education that left him unconnected and unprofessional upon graduating.

Since he bagn working at Griffin Cleft. I spend most of my day teaching him how to assist me and a good assistant should already know. Things like, when your boss's door is closed, you only knock on it in an emergency. Anything else can wait until he opens. It's a pain in the ass when he interrupts my artistic train of thought. Then, I have to try to recreate the variables that provoked the thought, once he's knocked on my door to let me know, that he left a message for someone.

All of my employees have to volunteer Two hours per week, helping children with their home work. How great would it be to have a finance guy join our efforts, and help the kids with their math? Because, secretly I can barely count. But, between me and you. Neither can Micah.

After letting Micah go, I went through Two more assitants in a month's timespan. Griffin Cleft is not rich enough to pay employees to learn. Since God keeps sending me lackluster assistance, we decide to introduce an internship program.

The program includes a Sixty day training and probationary period where the employee would prepared to take a salaried, general management role heading one of the companies that Griffin Cleft will launch in the Blue period. One of my first hired interns, was another young Black Man that I knew from the neighborhood. Marcus not only reminded me of what my nephew could have grown up to looked like. With the name Marcus, God had naturally positioned us to assume a uncle, nephew relationship.

It wasn't so odd that Griffin Cleft kept attracting true believers. I know it would be illegal for me to inquire about the subject matter. But, all the employees spoke to me about their strong Christian convictions, as though I wore a clergyman colar around my neck. And, Marcus was no different. There were days when I really admire how compassionate, and wise he seemed most of the time.

I could tell that he was annointed as well, but not sure that he had been chosen. My Mom always told me that I had God's annointing. I believed her. But, earning God's annointing is much different than earning his choice.

Like our mission for Classical Ballet. A component to Griffin Cleft's mission is to introduce Black people to unusual careers in the Entertainment Industry. I was lucky enough to land a spot dancing in Paramount Pictures, *Dreamgirls*. Looking at the performers on the set, you saw only Black people.

However, looking at the camera and the crew behind it, you did not see a single Black person, accept for Sharen Davis who designed our costumes. In showbusiness, there are so many jobs that are exciting, well compensated, greatly incentivised, and don't require the need for a Four year degree. These jobs are not as competitive as the performing jobs that require a great deal of talent, and luck.

I put Marcus to work on the Mad Science project. The first career I introduced to him was segment producer. I was reluctant to have him work on the project at first, because I didn't know how he would react to me having God as the Director of the project. It wasn't such an odd idea to him. With all of the autible clicks and bells going off in my apartment, he could tell that we weren't working without true direction. In fact, each time God would send us a signal that noted the correctness of a shot. It validated our work efforts, and that God was real, for Marcus.

Trying to catch up on some rest before Jesus gets back. He said he would be back in Six weeks, and I expect it wont be to crawl into bed with me. I've missed several missed calls from my Mom telling me that Uncle Gordon had been admitted into the hospital in Staten Island. I can barely here her messages because the planes are flying so low. And, so many of them flying out of La Guardia Airport. I mean Thousands of them, and with hardly any space in between their take off times.

Kennedy is the major Airport for New York City, and I don't know La Guardia to have such heavy departure schedule. And, there are only outbound flights. I think that it's time for uncle Gordon to become my left side.

Before I call Mom back, I take a minute to collect myself. I have a short conversation with God, and a quick prayer for my uncle. In the hopes that there would still be time to save his life. There is not much information that Mom can offer me when I speak to her over the phone as to his condition. But, she made it seemed as though the family should come and say their goodbyes.

When I hang up, I call his Two daughters to let them know they should head to New York to be with Uncle Gordon. One lives in Pennsylvania, while the other lives in Virgina. I don't think I got enough rest to give the situatuation the delicate passion that Mom deserved. I don't do well with dealing with death, so I ask Marcus if he would come with me. I'm glad he said yes, since he seems like family, and probably would provoke the same feelings from the rest of my family.

I stopped at the corner store to get a bunch of junk food, and drinks for us, before my driver grabbed me. It was completely dry outside, but by the time we picked up Marcus, which lives only a few short blocks from where I live. It began to rain perfusely. I mean, rain like a tsunami in the middle of the ocean. Everyone on the streets looked up at the sky in amazement.

My driver told me that people had been discussing the strangeness in how the planes were flying so low in the streets all day, while focusing up at the sky. All I could think is that, this shit is really happening. They weren't playing. "The planes are rescuing the One Percent first." I hear God explain to me. Right then, Marcus turned and looked at me with a look on his face that made it quite clear that he had just heard God too.

I give him a look back that hints at, I told you so. There is understanding God. There is Feeling God. But, not much in life can compare you to actually hear his voice. Crystal clear as if it is piped into ear buds, and Autotuned to perfect elocution. "I'm going to have to go through Jersey."

My driver looks back and tells us. "What's going on? The hospital is very close to The Verazano Bridge." I tell him. "All the tunnels and bridges into Brooklyn are closed." Driver replies.

As he turns the radio on. The D.J. announces all kinds of horrifc traffic detours, closures, and traffic jams. The wind is the stronger. The rain pours thicker. Marcus and I fasten our seat belts as we take flight out of New York City and the Old World.

As we exit onto the I95, and off the George Washington Bridge, Marcus and I pop our heads back to look at the apocoliptic sites. It appeared to be what I would equate the Forty day storm that Noah survived with in his Arc, would have looked like. Bridges and the Upper West Side of new york, being swallowed up by rising water.

My driver is transporting us so quickly, that it appears as though we are flying through the destruction we left behind. Once we arrive in Staten Island, I fail to get my driver to follow directions I know will be the fastest route to the hospital. But, he insists on following his GPS directions that take us through the scenic route that is riddled with traffic lights.

The sweetest sensation of love, protection and appreciation overwhelms me as I look over at Marcus. I look over and he is just chomping away at the snacks. Smiling, and so well adjusted to the situation at hand. Just happy and secure. Like a little boy, who I can't stop looking at with the perception that he is my young nephew returned to me.

When I ask him if he knew what was happening, he replied very candidly. "Apocolipse." And of course smiled. He didn't ever let on that he had been chosen to know the secret as well. But, here he sits with no question or doubt that he posesed God's annointing, and complete faith that he would be all good in the end.

Finally, my driver stops. "My GPS says that we're here, but I don't see the hospital." He says to me, as he anxiously clicks away at buttons on his GPS screen. I begin to tell him that it is not where he has stopped us. He ignores me because he is certain that the little box knows more about the town that I grew up in than, I do.

"We're no where near the fucking hospital!" I scream out at him to get his attention. "Don't fucking yell at me." He yells back. I fire him right there on the spot. Marcus and I hop out to find an alternative to getting to Uncle Gordon and Mom.

As he pulls off, Marcus and I head up the next busy road to see if we can hail a cab. When we arrived in the wrong location, I knew right away that his GPS wouldn't find the correct location of the Hospital. That GPS was programed to find locations existing in the Old World. The New World that we had arrived in was full of darkness and void. Just like God said the beginning would be.

But, the storm had cleared. As Marcus and I, walked upward towards Hylan Boulevard we felt at home, and I could feel that his energy matched mine. Eager. Naïve. But, excited. Totally the opposite of what you'd anticipate Two young men to feel as they approached what may be the witnessing of Uncle Gordon's death. I asked Marcus if he knew where we were. He turned and looked at me. With that heart melting smile, while shaking his up and down to signal that he did. "Saturn."

Clearly They have been talking to Mr. Marcus, as well. I pick his brain a bit as we continue strolling upward. Curious to know how much he knows. Our office chats have yielded more toward the Christian doctrine as we accept it to be on Earth. But, I thought I was the only one who imagined that Martians were believers in Christ too.

We arrive at our major crossroad. Where Clove Road meets Highland Boulevard, there is the Staten Island expressway right in the center. It is sandwiched in between Two service roads called Gannon, with one directional traffic either driving North or South. It's a very large thoroughfare that divides Eastern and Northern sides of Staten Island.

I thought for sure that there would be cabs for us to hail. But, there doesn't seem to be a single car driving on the roads. There is traffic on the Expressway, but who's hailing a cab there? "Find your home, and you will find us waiting inside." I hear God say. "I know where it is." Marcus teases with me.

Ok, being the skeptic that Mom raised me to be, I think "What the hell is he talking about? I live in the Bronx!" I look to see what Marcus thought, and couldn't find him. There I am spinning on a gorgeous, dark street, lined with willow trees that breaks the moonlight up like a fairytale drawing. Than I find him standing in a yard windmilling an arm to tell me come join.

I arrive to find that he is actually standing in a driveway that splits between a small White, wooden home, kind of like a cottage. To the right of it is a much taller home built out of bricks, and great mason work. Parked in front of one is a brand new, Black Hummer SUV. In front of the other is an odd looking, White old school, funkadelic mobile home. "You have to mark your territory." Marcus offers me with a smile. "Who are you?" I ask him. "I'm your Navigator." Marcus replies. "Chose a home for yourself, and one for Marcus." God tells me.

I than take out my dick and start spinning in a circle as I took a piss on to the front lawn of the bigger home. Marcus, dodging my sprinkles, yells at me "What the fuck is wrong with you?" "I'm sorry. I didn't know what you meant by mark my territory." I apologize. "The keys are in your car." God tells me. I don't ask which one, because we had already discussed me driving this mobile home monstrosoty through different quadrines.

It's main purpose was to include hightech surveilence inside, that would monitor the children of this world. But, he wanted the outside to be nonthreatening to the kids, and would be cartoonishly inviting. I would be the only adult the children would be in contact with, as this world would be where the children were sent for boarding education.

If memory serves me correctily, there should be no keys. In this world nothing locks. "You're not there yet." God tells me. I walk over to the White, funky mobile home and pull the rear door, but it doesn't open. Confused, and eager. I run to driver side door, but it's locked as well. Then I remember. Both of the vehicles are mine, since Marcus is my navigator. He'll always be with me.

The Hummer is for when I am at home, not at work. But when I try to open the doors to the Hummer they are locked as well.

Marcus walks into the street to tell me, "Maybe the doors are open." I run up the stairs to the brick house that are locked tightly. Marcus and I meet back in the driveway. "Mine is locked too." He tells me. "Find your home and your family will be in it waiting for you." God tells me again. "God I don't understand what you mean?" I ask him. "For every clue that I give you. You will have to let something on your person go." God replies.

"Marcus hand me that bag of snacks." It's full of stuff to let go. Thankfully, because I am clueless. I take out a can of Arizona Tea and trow it into the street. "Ok. Now, give me a clue. Please." I say as I look up at the sky. "Marcus is your navigator and is rquired to help you."

What the hell kind of clue is that? When God refuses to help me any farther, I turn to Marcus. "What is he talking about?" I ask Marcus. "I need my GPS but my phone is not getting service here." He replies. "Oooh! That's why God had me buy in Five phones. I have a lot of your stuff." So, Idig into my cute, Tan, leather shoulder bag to pull out his brand new Lumia. "Here, I think this phone belongs to you."

Marcus grabs the phone and powers it up, as we start walking back to Clove Road. I guess we're supposed to go to my family's house. But, I thought we'd all be traveling together. I'm not sure where they're settled in. We turn on to Clove Road traveling Southbound. Still not a single car on the road. It's unusual, because this is a popular way to get in, and out of Brooklyn.

Then I remember that the Bridge is closed. We stop in front of firehouse that looks like Snoopy and the gang should all be hangging out front of it. Everything looks overly exaggerated and quite cartoonish. "You have to go into the Vortex and pull the chord that unlocks everything." Marcus tells me. "Where's the Vortex?" I scream out.

I'm getting very annoyed at this point. God then tells me. "You know the rule...." I take a bag of chips and toss them to the curb. "You're looking right at it."

In the middle of the road, and directly across the street from the Peanuts fire house was a large, round, patch of beautiful green foilage. The foilage made a cave that circled around a seating area were marble benches lined the inner ring of the bushes and trees that feel over it, at top. In the very center stood a tall flag pole, that had no flag raised. So, I assumed that raising the flag is what Marcus means.

"Come on Marcus, let's go home." I tell him as I run across the street to the Wonderland flagpole. "I can't go in with you." He tells me, standing behind in the firehouse driveway. "Why not?" I ask him. "Because, I am a kid." He replies. He's nearly Twenty Four, I think to myself as I strug my shoulders, and enter the Vortex. Inside the Vortex, I can't stand up completely. So, I run around the circular path, hunched over before I take a seat on one of the marble benches.

I take a moment to soak in the beauty. A solice moment asking God if this is really happening. I also wonder about the chaotic world we'd just escaped. I wondered if people are dying at sworms at a time. I look around to see how you raise the flag, or indication of what to do, and find nothing. I need to see if Marcus has instructions. "What do I do?" I yell out at him. When I don't get an answer I run out to check on him. "Where the hell did he go?"

I see a flash of light in the sky. "No way! Did Marcus just fly up to the sky." Wondering how he did, I take off running North on Clove Road hopping to gain enough speed to take off. I run. And, I run. Then I jump towards the dark sky thinking that I would begin flying. As I return to the ground beneath, I wonder why that didn't work.

I have a replica of Marcus' phone that I pull out to see if I could communicate to Marcus. I press the lock screen, thinking that our phones were masked as the phone. But, were coded with functions specifically for God's mission. So, I press the lock buttton, while speaking into the microphone. "Marcus. This is Mark. Do you copy?"

I don't know. It's what they always say in the movies. "Marcus. This is Uncle Mark. Do you copy." I say again. But, still nothing. Then I see him turning the corner from Hylan Boulevard. I run up to him. "Both houses are still locked." Marcus tells me. "I couldn't figure out how to raise the flagpole. What does the instructions say?" I ask. "I've been asking God, but seems like he only answers your questions." He replies.

Oh. "I forgot to tell you to throw something away." I say. "I was wondering why, you through my favourite chips away." He circles around himself, screaming at the sky and beating his fists on his thighs. "I'm over this bullshit. I need to get back to the Bronx. Can I have some money for a cab?" Let's find an ATM and I will get you home.

Marcus then turns and storms down North Clove Road. I follow him thinking that there would be a gas station just a few blocks away where we could get some cash. "I don't see it." I whisper in shock. "Mark, stop playing with me!" Marcus screams back at me. I'm not kidding you. I've past this corner Thousands of times in my life. There was just a gas station right here a week ago. Now it's completely gone. "Fuck this. There's a Paramedic right there. I'm gonna ask him how to get back to ferry. Then I will find my own way back." He yells while crossing the street. No more of that sweet smiling. Though I admit even in his rage. He's still pretty adorable.

As I head South and upward on Clove Road, I realize that I let Marcus leave with the navigator phone. I have only the empty plastic bag left. So I toss it, to get another clue from my father in Heaven. "Remember the 60's." God is talking about the 1960's, Black Civil Rights Movement. I've often told him that I feel as though I missed the greatest time to exist as a Black person.

When creating comodoties for a new nation, like I would be doing for my new world. Leaders have to work in professions outside their primary focus, just because there is so much to get done. So with this clue, I begin trying to imagine my life at its most rudementary existance.

I look to my right down Hylan Boulevard. I see that the street is lined with work Vans. I read the job titled on the backs and sides of the vans to see which blue collar job I'd be ok with living with. Since contracting has a great deal of interior design inclusion, I choose to be a contractor. Once I make my choice, I am shown an imaginitive visual of what my life would look like. My husband and I live in the front of the house, and we walk around to the left of the house where our parents would be living in the side of the same house.

I set off to find the house in the visual. At this time it begins to rain just a bit. Still walking upward and South, however where I am now has not been completely duplicated yet. There is only the Staten Island Expressway, and the service roads, which are being constructed. I'm not entering expressway traffic so, I decide to venture through the hazardous construction zone. There are big trenchs with gapping holes in the ground. Hills of concrete rocks, and caution tape, construction trucks, but I'm alone. Well God and I are alone, I suppose.

After checking a few neighborhods, I give up. "what am I doing here God? This is pointless." I hear nothing. I only have my wallet and keys on me. I take my key ring out, and pull off one of those pharmacy discount club cards. Spiral it off, and drop it onto the ground. Then God speaks, "You must choose either to go to your Mother's House. Or, your Father's House."

By Father, he can only mean one thing. Go to my home church in Brooklyn, or that he's going to kill me so that I can join him in Heaven. Not really wanting any of those Fatherly options, I continue moving forward towards where my Mom lived in the Old World.

As I tag along the extremeties of the Staten Island Expressway, I take a look at the traffic driving North. I notice that there are only about Twenty different variations of traveling vehicles. I especially notice that the charter bus line is one that I have not heard of before. Yet, their driving by me every Four minutes. Electiric company trucks too.

Odd, but the traffic seems to be on a loop or something. I then jump over the boarder, which happens to be a Fifteen feet drop down onto the Expressway. Wanting to prove that the traffic was looping, I zig zag, in and out of traffic. Timed specifically to when I thought the traffic would stop.

As I walked further South on the Expressway. New loops would unveil. Now with an angry driver throwing road rage at me. Amazing, the same Red, 90's looking , 3-door, Red Corolla that slows down, stops, ye'lls, "Get outta the way asshole!" And, then speeds off again, happening about every Six minutes.

I feel as though I have been placed into a video game. I get off the freeway and see that emergency assitance vehicles are idol. Ambulances, Fire Trucks, Power Trucks, all parked along the street. I didn't want to be seen by anyone. So I dip into a very familiar short cut I've used to go to Mom's house many times.

As I walk through, the quiet, beautiful residential neighborhood. I notice that the street signs were names that had significant relation to my child hood. And, to my future story as I know it to be. Like a street named after Diana's youngest son. All of sudden, nothing looks familiar. Accept, the fact that I have been walking in circles for about Twenty minutes. I walk to what I thought would be the main road that I dipped out of, a minute ago. It wasn't. The rain has begun increasing, while lightening also starts to strike. This is rediculous. I spiral off another discount club card from my key ring. "Follow me." I hear God say. My inner GPS fires up, and my body takes off, mazing through the streets to get me back to familiar land. Cool, I know where I am now. Mom's house is just up the road.

Opened the gate, run up the few steps that lead me to bangging on her door. I see that the lights are on, but get no answer. I bang and I bang, hoping that she is just inside and asleep. But, I give up eventually. I sit on her porch for a second, since it's covered. It's still raining and it gets heavier by the minute. I spiral my mailbox key off of the ring, and toss it while asking God for another clue.

"You have only Two hours left to get to either your Mother's house or your Father's house." God tells me. If I am already at my Mother's house, I'm not sure where else to go. Each time lightening strikes, I fear that it's aiming for me. Since he has shown up, I feel like God has been wanting to show me of how harmless death is.

I run to ring Mom's neighbors to see if they at least were still the same famalies, in this New World. Funny how One only speaks Italian. While the other speaks only Russian. I learned some Russian while moonlighting as a dancer in a Russian Club in Brooklyn. So I chose the Baryshnikov family.

The Father pokes his head out of the second floor window first. I can barely make his face out as the rain pours into my eyes when I look up. But, I can hear him loud in clear. In every language, Nigger sounds the same. He hates Blacks, especially their Sons.

Communicating in unbelievably perfect Russian, I am able to ask of Mom's where abouts to the wife who has now poked her head out. The husband is only concerned with calling the cops because I rang their bell so late. As if he wouldn't do the same thing if he worried about his Mom. The wife tells me that my Mom was taken to the hospital. Not knowing what happened to her I panic. I ask the son to call me a cab, since he has now poked his head out of the window. He does, and I sit out on the curb. In the rain, awaiting the cab to arrive.

On the way to the hospital, I ask the cab driver to stop at a Chase bank. When we get there, I remember that I had thrown my Chase card away to get a clue earlier, from God. But, I hung onto my B of A card, which for some reason is now being declined entry into the Chase branch.

WTF! I've never heard of a Bank denying access to competing bank card holders before. I hop back in the cab and ask the driver to take me to a B of A that's on our way. He starts off, and we have a very nice conversation getting to know one another. All of sudden, he flips and starts telling me that he wasn't going to be wasting time looking for a B of A.

Instead he planned to take me to the next gas station ATM. I take the remaining cards from my wallet, and drop it on the floor of the cab. "Clue Please?" I offer up to God. "Jump out of the Cab, tuck and roll, and run the opposite direction that the cab is driving in Three, Two, One, Now!" He tells me, as I open the back door doing exactly as he says. I run a couple of blocks, noticing that there is a 7Eleven that is open.

Aside from the Emergency responders, the roadrager on the expressway. This is the first time that I have come in contact with pedestrians on Saturn. I walk through the store, checking to see that all the commercial brands are still stocked the same. By all appearances I haven't gone anywhere. I get a couple of waters and head out of there.

I need to hang onto the cards and key that have remaining. So, I toss one of my cell phones. The battery on both of them has died at this point, and pretty useless anyway. "Walk straight towards the Beach." God tells me.

Each time lightening strikes, I duck behind a tree. I know that I have only minutes left to get to Mom, or my Father is going to take me with him to Heaven.

I arrive at the Hospital finally, and "You can't go in that way." God tells me. "How am I supposed to go in?" I ask. He doesn't answer me. All I have left to leave behind now is the shirt that I am werating. So I pull it off and toss it onto the hospital lawn. "You have to enter through the Vortex." He tells me. Here we go again with nonsense. He leads me through the vortex, but not before giving up the use of my eyesight, and full use of speech.

By the time I am at reception. I can only speak letters, not allowing me to use the letters again once I have spoken them. Communicating with hand gestures to people that I can't even look at, I get them to conclude that I have had a bad reaction to drugs.

Knowing that I did not take any drugs, God tells me to just play along. He is insistant that I should not open my eyes, because he is among the entourage that is wheeling me back to admmitance. As they try to figure out what I had taken, I hear Mom's voice. "You know this person?" I hear a nurse ask Mom. "Yes, he's my son." She replies. "Mark, honey. What happened?" She asks me. I want to answer her so badly, but I'm still under God's spell that has me blind and mute at the moment.

When my wheel chair is stopped, I feel hands grab me to help me onto a gurney. "Listen to the females voice. Only respond when you feel these Two taps on your right shoulder. Can you feel that?" He tells me. "Yes I can." I say to him in my head. "That's me. But, keep your eyes closed, because I don't want you to see me." Then the nurses contnue speaking to me.

Now, with my eyes closed, light begins to part the darkness from the shade of my eye lids. I did not open them, but, the vision made it seem as though I was looking at something. Something marvelously beautiful. Me standing on a clift in front of the sun. Puts you in the mind of Heaven.

The nurse tells me that I am going to feel a pinch. But, it's only an I. V. Scared of needles, that gets me to open my eyes. I look around at all the faces hovering over me. Mom, a couple of nurses, a couple of security guards, and then the same cop that took me to Lincoln Hospital when I got arrested. Before I could say God it's you. I passed out cold.

When I wake up, all I could think about was what they shot into me. I saw the needle, and it was not an I.V. I can't help but notice that there is no I.V. in me now either. However, when I ask Mom about it. She claims that it and there was an I.V. "You want to go say hell to Uncle Gordon?" She asks me. "Of course." I tell her. On the walk to his room, I think that I completely forgot that Uncle Gordon was the entire reason that I embarked on last night's journey.

Am I going crazy? I wonder. What the hell was last night all about? After spending a little more time with Uncle Gordon, and Mom. I grab my discharge papers and head back home, to the Bronx. Home to now spend more money, and effort regaining the items I tossed during God's Great Goose Chase.

When I get home I can feel the presence of Jesus. He has returned after being absent for Six Weeks. I can't deal with him right now. So I go to sleep. Sleeping all evening, through the night and awakening the next morning.

I turn on the Television when I first wake up. The screensaver on the Smart T.V. has bubbles bouncing all around it. How cute? I didn't know that I could change my screen savers like that. Why has it not done this before. I never believe that I am completely alone. But, today for some reason I can feel foot steps on the floor.

When I crawl into bed. I can feel the bed shifting, like someone is getting into bed beside me. When I go into my studio, I can feel foot steps follow me into it as well. Odd, I wonder who could be so present in my home with me. Jesus starts to explain to me what has happened.

He tells me that Mom has died in the old world. That my family is no longer the family that I knew. "In a since you've been cloned, and will replace the Mark that this family knows." Jesus spoke, and continues. "No one ever dies. You simply transition from one world to the other. You chose your Mother's house. Therefore, you went with your Mother."

"Am I in Heaven?" I ask. "No, I have chosen you for a very specific purpose. I am your Lord and Savior. You shall live eternally." Jesus rewards me with. However, I am unimpressed. Every since my Christ family has arrived, I have been played left and right.

There is not a day that passes that I don't wonder if I am being punished as to rewarded. It seems like one big joke to see how crazy a perfectly sane Human can be driven to insanity. "Stop playing in my hair!" Whatever this is in my house, it's getting on my last nerve playing in my hair. I hope this isn't what God meant when he promised that I would see my little nephew, Marcus again. Which reminds me, I got to call Big Marcus, to see if he made it home ok.

Since he didn't answer, I went over to his house. I spoke with him and apologised to him. He forgave me, but presented me with a perception where I am the only one who went crazy. Granted, running, trying to fly to Saturn is plum fucking nuts! But, he was right there witnessing bridges being gulfed up, and claiming to be my navigator. At any case. I'm glad it's behind us. I go back home to get some work done. Not before stopping to get a new phone.

The first person to ring me up on my new phone was Mrs. Four Octives. Right on time too. Because I need another True Believer to vent to, about the shit that God and his son are taking me through. "I'm sure you've heard the news by now." She tells me. "Girl, you know I'm out the loop. What news?" I reply and ask her back.

"Jeremiah passed earlier today." She explains, as my heart falls to my feet. I'm not surprised and yet I am shocked beyond words. Not surprised because if I were God. And, had to combine Two people into One. If I were already chosen, I couldn't imagine any other person in the world that would complete the other half, besides Jeremiah.

When I hang up the phone, I realize that the feet steps that I feel, and the person playing and my hair. And, the woo, woo, woo, it's okay ora that is dominate in present. Must be Jeremiah, trying to console me. Console me? I don't even want to think about what this means for his mom. He and I sit down and write a heartfelt letter, to reach out to his Mom.

I'm not sure what to write, since she only has Jeremiahs last impression of me, and has not heard my side in years. I include things like You still have a son left here wih me, and I hear Jeremiah tell me. Don't say this. Don't say that. It's too long. You're saying too much. And so on.

Eventually, we get the letter off, via e-mail. We're not given much time to chat or catch up. "We're going to go now, but we're coming back." Jesus tells me. Then I feel them leave like air that escapes through a cracked window.

Chapter 11: Coming and Going

Here's where I fuck up. Unless you buy that whole idea, I can't get it wrong, and I can't get it right. Or, perhaps it was God who fucked up. I don't know. It may have been Jesus. Why would he tell me to study the number Six, for Six weeks? Knowing as well as I do, that anyone would have thought it were time to study the origin of Satan, with that arrangement.

Am I alone? One of the things, that didn't make me a true Believer, is that I didn't believe in a Hell, or a Satan, or a Devil. I just thought them to be clever ways for the Church to collect tithes, and enforce religious standards over nations. But, in the past six weeks, I experienced the company of a scary being.

The craziest night with it, had me in the studio spray painting letters onto the walls. When I felt that I had gone too far, venturing into perhaps Devil Worship. I rushed to cover the letters. Then as the letters disappeared behind the coats of paint, that I rolled onto them. I felt as though I was erasing meaningful initials, or abreviations for something. So I left one letter uncovered. The Letter J. I thought that it could stand for Jesus, and certainly didn't want to erase him from my life. After only just getting him back? There is still so much left for him to teach me. But, now I wonder if They mistook the letter to represent Jeremiah.

When I adopt a new rule, or whatever to my belief system. I rarely challenge the belief. I know that whatever is not true, can become true with a little belief that it will.

I only question that letter J representing Jeremiah, because I've an abundance of evidence that my thoughts, ideas, and dreams aren't my own, entirely. And, that pieces of my Purpose-Puzzle are giving to me over a lifespan to paint specific, more complete pictures when the fullness of the Purpose culminates. For more than Five years. I have been planning my transition from dancer to whatever will be next. I knew that I would change my name. Firstly, because the odd spelling of Omar, is a pain in the neck to get people to spell correctly.

Secondly, because I wanted to have a name that lacked ethnic relation. But, when God gave me the name Mark. He and I also planned to use the Coming Out Celebration for Griffin Cleft, as the event to announce the name change to the world. Our plan was to title our first press release, "A Gay Man. Commits Suicide in Los Angeles, and lives to talk about it in The Bronx!" The press release would go on to include a small story of how Omhmar, became an entirely different person, by becoming Mark Griffin.

I would delete all of Omhmar's social media after a Thrity-day memorial period, where Omhmar's social media followers could say goodbye to Omhmar. The press release would conclude with the line up of the Griffin Cleft Launch.

Now that Jesus and God have chosen, trained, and prepared me for Jermiah's arrival. I wonder if They new all along that the press release would expand beyond my initial intent. More importantly, did I have something to do with Jeremiah dying? I'm familiar with close family dying, almost ritualisticly, when someone is accepted into the Illuminati. But, I'm not quite sure still what is going on. Jesus assures me that we're doing something that has never been done before. So how could any of this be Illuminati, or anything else that has ever existed?

It will be a physical event. That's what I've heard them tell me. And, that is what I believe. Yet, to be winning, it feels more like loosing. Each day I loose more happiness. More focus, more passion. I ache with the loss of friendship, companionship, and finances. I'm ready to go, and yet hendered from moving forward. I've never wanted to stop. But, it was in my design for Omhmar to terminate his lease of the body, so that Mark could become its new owner.

In order for the big switch to take place, Mark had to let go of all that was Omhmar. Leaving him to be the rock bottom, lump of clay that the Christ Family would have to reshape back into something yielding of success. After being completely obedient, all the while standing in my truth, task, after task. Event after event. Person in front of person. And, they all tell me that I succeed. I impress. I surprise, inspire, and am wanted. But, I still loose with each passing day. Worse of all, I loose more faith. More belief. It sucks that when you are chosen to back up, both, Jesus and God in competition. No matter who of them, wins. You loose.

"What does Our Father In Heaven expect from a Man?" Jesus asks me. "Decisiveness, Consistancy, and Strength." I reply back. This is the second day of Jeremiah's transition. Jesus has returned to me, without Jeremiah. He begins with a quick review of the first quarter of my studies and participations.

He also reminds me of which callings I've agreed to accept from God. Teaching, Prophecy, and Government Leadership. "Like all ancient Seers, like Nostradamus. You have just had your Psychic Initiation." Jesus tells me. "For your next quarter, we'll begin exploring ways that you and Griffin Cleft can teach and lead." He continues. "Spend time this week, re-imagining." Jesus commands, as he makes his exit.

If Jesus is supposed to be the practical one, he should stop immitating his Father's mysterious ways. Re-imagine what, I wonder? "Start with Hotel Nativity." I hear God speak to me. "Oh good, you are still here." I reply to him. "I'm always here." God says to me. "Why did Jeremiah have to die?" I ask him. "He did not die, you did. Remember?" God asks me.

Yes I remember what Jesus told me about coming with my mother. But, come on! Mrs. Four Octives called me, herself. And, the news of Jeremiah passing is all over the internet. The children are devistated by his home going. The Holy spirit comes over me as I mull all this over in my head. When it does, I grab my phone to look up the origin of the Number Six.

There it is in plain broad day. God's number of completion. I don't know why I always rememberd Seven to be God's number of completion. But, this news puts a great deal into perspective. I often can't tell when I am supposed to be acting for practice or for purpose. I guess it should require some seriousness, providing the context of God's process. However, I can imagine, that having made it to Number Six, in our exploration of numbers. I have completed enough of his process, to start the purpose of God gifting me with it.

"Omhmar + Jeremiah = Mark," God informs me, and continues with. "Jeremiah has agreed. So has, Mark. It is now, Omhmar's turn to agree before you all can continue forward."

Which came first? Was it The Chicken, or the Egg? I suppose it's a question of perception matched within your character's disposition. When ever I perceive the beginning of manifestation. I often think that there is a clear answer, apart from perception. The Chicken came first. And, of course the Chicken was Female as she was introduced, and with Chick.

Any true believer understands this to be factual, as God uses the Imaculate Conception of Jesus Christ not only providing us with the Biblical source, of our Lord and Savior. But, also to provide his children with the scientific explanation of how Humans came to exist on Earth.

Imagine that we are all in one achord. And, we accept that through millions of years, cells, intelligence, and the physical make up of our Human bodies, evolved to the physicality that is the Homosapien. The world where we orignated, was destroyed by God while he inplanted us all in another world to continue Human evolution. God then, inserted a Woman, with male child in her belly, on to The Earth we are modernly familiar with. Once the son is born, he and his Mom commit the acts that procreate our exitance. Leading us to eventually inhabit the Earth, and Heavens. That's some relation there. Talk about accepting a calling. I think that our story would be quite short, if I were given the task of literally being my Motherfucker.

You could also take the result from our Chicken/Egg debate, and assume it to be eveidence of America's most defined cliché'..... Ladies First!

My Christ family has got me so busy, that I don't really have the time to mourn the loss of Jeremiah. Not, to mention that it feels less like a loss to me, right now. And, more like I got him back. Remember that, before his transition. Jeremiah wanted nothing to do with me. I guess the jokes on him now. He probably died holding onto resentment for me. And, now will be forced to live in my body. He did say that Revenge is Mine, sayath The Lord!

Please don't read the telling of my voyage, to be one that has a purpose, clearly defined, budget, itenerized, and staffed with navigators and ship crewmates. There are financial struggles to say the least. There is a challenge of not knowing what I am getting myself into. And, trusting the odd process that is full of gifts, and messages that makes no since to me. Yet my faith and God, is what keeps me going. I have no clue as to what physical event will allow Jeremiah to dwell in the left side of my body. But, I have always believed that this is the event to expect.

There are very little differences that allow me to know that I am in fact in different worlds. When God guides my transistions to them. And, I am now in my Fourth World. Other than, my faith to believe almost everything God tells me. He did inform me that when chaneling him, he doesn't provide me with any clue that he is not among the other voices that I hear.

Which makes it nessacary for me to doubt his word at times. He has demanded that I treat him the same as I treat the others. And, that I will never give him any more trust than I give the others. I don't trust the others, at all. So I see why the Bible says that you only need the amount of faith equal to that of a Mustard Seed.

I think that I really wasn't taking this process and physical event serious. Hard to say. I've provided you with much eveidence of how I did in fact believe it seriously enough to participate in. However, Jeremiah actually transitioning to the place where he's basically a Magician. Has me confused about what our specific purpose will be. I can't imagine why God would have taken a life, just so that I could tap into the Grand Humanity that was Jeremiah Kyle Tatum.

He was his Mom's only son. She is still not answering her phone. But, I keep sending her emails, praying that she gets to read them. I have every intention of moving back West when my lease is done. Or, maybe even sooner. While completing the desing of my home/office, I was in the belief that I was not designing the place for myself.

But, for someone else to perhaps live in it, holding on to the lease, while I lay roots back West. It's very popular for residents with Rent Stability in New York City. Originally, I thought it would be my older sister. As she would have to hide out here when God returned her son to us. But, hopefully it was all for Jeremiah's Mom.

With every stich of Los Angeles holding siginifgance to the life that she and Jeremiah once shared. I think that moving to New York. A City that lacks Jeremiah familiarity in it. Could be just the move that would help her mourn, but move on to a new life lacking of him. None the less, I knew it would be a woman.

Since Jeremiah was the head of his household. I was clear that his Mom was going to need some help replacing the income that supported them. So, I took the liberty of setting up an Artists Management company for her to run once she got here, to New York. It was a company that I had already done the development work for. Back in Los Angeles, when they asked me to create the Management company the first time.

MaWest was an excellent stage mom to Jeremiah, and I've always thought she could get rich off of guiding other kids through Show Business. Instead of applying all of the transeferable knowledge she picked up along the years, to taking the company to scale. She, did nothing with my year worth of hard work.

Maybe this time, she'll give it a better crack. This week, our priority is Gods party at Griffin Cleft. Until now, I've been pretty silent as to me being chosen, to my family and friends. Not for any other reason than me being so busy, that I forget to talk about it, when I see them. I'm telling you, working with God by your side will wear you ass out!

There are times when I beg him, "Please Dad! Can I go to bed?" And, not even a bed. Can I get a moment to myself on this here park bench to catch a couple of zzz's. Sometimes I will be riding the subway, thinking. What should be a thought to myself. And, then God will pop in, to add his Two cents to my idea. Come on, underground? You're still there hangging on to every word that I think. And, intruding to correct my disbelief. Which can often seem insulting and obtrusive.

Through out the fall we have several parties planned for the purpose of filming some lively events to make Mad Science most interesting for our viewers. As we end the summer, God and I planned a Griffin Cleft event at my home. Mostly for family. The idea behind the party is to introduce the idea of God working directly in my life, or that I have been chosen. Not in a gloating kind of, Ha Ha, I'm chosing kind of way. But, to speak about Same Sex Marriage.

Since I submit to God's will, and allow him to chose the husband that I will marry. I'm not sure what the problem was with Officer Usher. He was quite phyne, and perfect for me. However, if he were truly a Cop, I may have reservastions to him. And, only God would know.

At this laxed event that we're throwing, my guests and I watch the movie, *Pride and Predjudice*. This is a movie that God led me to, in effort of reminding me what marriage is truly purposed for. And, how the original ritual came about. I'm not sure if God has the plan to include Mom, or any of my other family in, on the search or decision. But, we both agree that this would be the best way to introduce to them, what's been going on with me.

My producing chops are more cursed than blessed at the moment. It's been a few years since I've produce anything. Omhmar has produced everything from Opera, and Ballet. To Reality Television and Video Games for Activision. But, this will be Marks first time producing anything, since he produced the Book of Mark, for The Holy Bible. Here now, since Mark has joined the living, he has had very little obligation, nor issues to make bad decisions in.

Many decisions are being made this week. Griffin Cleft letter head, plaques that label the rooms of the office. Note cards that allow us to announce free tutoring to the neighborhood children, along with God is Good! Not to mention that I still posess the problem of not being able to do anything in a small way.

I have to go above and beyond expecations of a single purpose. So for this party, I am doing the most. Including a promo video of what to expect from the launch of Griffin Cleft, and I have a singer friend, performing one of the musical selections from Hotel Nativity. Which will be our first movie, once we arrive at the place where we add film production to Griffin Cleft.

This morning's staff meeting is off to a late start, as usual. The only thing harder than getting these Bueens to come in on time, is getting them all in the office at the same time. If one isn't calling in sick, another is being arrested by NYPD, for some bullshit. I'll spare you the names of my ratchet interns. Aside from Marcus, I call one Wicked Bitch One, and the other. Wicked Bitch Two. Hereon out, referred to as WB1 and WB2.

We're an office full of fatherless children. Raised soley by our single Mothers, and packaged with issues resulting from the single parent household. Days where we are really tapped into, the work and mission that we're doing, definitely rock! However, most days, they look for me to replace the love and fatherly compassion that we've all been without. It can make it hard for me to be a boss, and enforce the law. Not to mention, that I am barely, just a few years older than they are.

The last thing I want to be to them is a father figure. Please at least let me make it to Forty before the kids start calling me Daddy. We're using the Griffin Cleft Internship program to double as an incubator for a new project that we'll introduce later, in Indiana. The project, when we arrive in Indiana will become an education system for young boys between the ages of Nine and Fifteen years old.

But, since NYC has rehabilitation programs offered through the Human Resources Administration. Jesus and I figured that the Internship would be a great way to introduce the idea of Autodiadect. Some might say that I lucked up in attracting Three interns who were participating in the HRA programs. I'd prefer to think it was his design all along.

My entire perception of the Purple Period, is that God and Jesus came to prepare me for fame and fortune. I always looked forward to the fortune.

But, not the fame part. They made it clear that Griffin Cleft should be based in One of America's smaller cities. Away from tourist attrcations, and asked where I would like to live.

Thanks to Dancing, I've seen the entire country Twice. And, of it. My favorite city is Indianapolis. A city where Black people can pay Two Hundred and Fifty Dollars for a Classical Gala in a packed Opera House. And, afterwards. Those same Black folks will serve and dine on Fried Chicken, and other Soul Food delicacies. They support the gay community to the tune of an Eight Day Annual Gay Pride Celebration. The longest in the country, yet allow their business owners to reject customers based on sexual orientation.

God has asked me to only study Numbers this year. He has specifically asked me not study Colors. Not, quite sure why not. But, asking me not, of course makes me wonder about their purposeful origin. It wouldn't matter to me much, if I didn't know that God doesn't create anything without it being the exact answer that resolves a conflict on Earth.

As I visit Mom for the first time, since arriving on Saturn, I am paying close attention to the environment of Staten Island. Everything looks the same as it did on Mars, and back on Earth. Accept that the Sun shines so much brighter now.

I assume it's because we're higher. I really don't know enough yet, to tell you the relation of the worlds, in terms of direction and distance to one another. It feels as though, with each transition we rise. In this world, there are times of the day we're the clouds are so low, you can see through and over them. The Sun glows like a spot light fixating you on its daze. And, as a result color seems to be magnified.

The skies look like water colors, while the trees seem to create a compact proscenium that's shiny, and vibrant new cars cruise through. As though Oprah gave the entire United States variations of 2015 Jeep Grand Cherokees. There doesn't seem to be an older car in sight.

Could it be that this is the first time I've actually looked at nature. Could I have really been so busy all of my life that I never even onced noticed, how beautiful the Sun and Moon actually are?

Mom seems normal. She could be a little shorter. But, I've heard that we do shrink as we get older. So funny, how I use to think she was so tall, when I was a child. And, now to tower over her the way that I do, makes me wonder. "How did this little bit, put the fear of Jesus in us, as kids?"

Uncle Gordon is out of the hospital, and doing much better. And, back to his crazy self, as Mom and I pass him, leaving for one of our walks. It became a thing that I did to help her stay fit when I lived with her. I still, usually make her walk when I visit. She's been waiting a long time for me to return to the Church, so that she and I can fellowship in his word. But, there is really only one thing I want her to share with me today. That is, what the hell happened to me when I was Seven years old?

I don't have many ways to validate the information that I hear. Mom claims that I don't remember lapsing time in my childhood, that corellates to the period that I am supposed to have been abducted by the Martians. She said very specifically when we last spoke about it, I don't remember where we lived when I was Seven years old. And, that it was too painful for her to talk about at that time. But, One day she would tell me.

Today she has no idea what I am talking about. Strange that I can remember exactly where I lived at ages Four, Five and Six. And, then again at age Eight through the present. No luck validating the abduction. But, at least we cleared one thing up. This person strolling through Wilowbrook with me right now. Can't be Mom.

God's party was disastrous, to say the least. Only my family showed up. Wit the acception of Mom, everyone else stayed on their mobile devices. It was an event of people listening to Mom and I talk. Way to go God. The incompotent interns that he sent me didn't even show up, to work the damn event.

WB1 popped in for a second, to pick up his stipend, and his third strike that eliminated him from the internship. Both Marcus and WB2 were no shows as well. But, we decided to keep them on for a bit longer.

The money that I spent on the party, could have been beter spent on taking Mom to dinner. I'm getting a little tired of always obeying God's every command. Well it's impossible to obey them all. Let's just say that I work my ass off for my CEO. And all he gives me in return is empty promises that he will replace whatever Jesus takes from me. Which is a lot.

When I agreed to plan my giving. We came up with a plan for which to dstribute philanthropy through Griffin Cleft. But, Griffin Cleft aint bringin in no money. Mom can foget about a car now. I've spent most of that down payment. Partly, because of our philanthropic motivation. And, also because I am feeling like no one in my new world gives a damn about me!

I try shifting my focus to my compositions. At first it was an attempt to do something that was Un-Cristian. If I'm not walking down St. Mary's Street, I'm looking out of my window to see Noah's Arc, and Mayflower moving trucks. Now, I look to some old works to reimagine. And, there both about Jesus as well. It was not my intention to become the next Tyler Perry.

But, it seems as though my creative mind only thinks Biblically. I start by looking at a choral work that I was composing for a possible requiem. I've often thought that there should be a requiem mass that emulates my perception of transitioning. There all so sad, and I thought that I could write a requiem that was bright, but respectful of the text.

I began working on it Two years ago, and now some how think that it could be something to complete in honor of Jeremiah. My focus shifts to a musical stage play, I wrote in 1999 called Hotel Nativity. I tried to prodcue it before leaving Los Angeles. I had an all star cast excited to perform the play, but. The development firm that I used fucked up their deals.

Which is actually the event that transpired my journey that creates Griffin Cleft. Although I have re-written Hotel Nativity six times, I have no intention of producing it. God, and I have come up with a brand and process that Two-Act Musicals don't fit within anymore. Now that I reimagine. Jesus is leading me to invision a film adaptation of it.

At today's board meeting, I decide to pitch Hotel Nativity to the Board as a film. Instead of a staged musical play. This is where you'd really think I were skitzo.

Every Monday, at noon. The Sun shines directly into the Four Octive Family Studio where I stand in its center. I believe that God, Jesus, Mother Mary, and the others hold seats on the Board of Griffin Cleft. I introduce new business as well as updates on the old business. I hear feedback, guidance, and we actually take votes. The meetings are filmed and archived.

For new business today, God introduces some old ideas. Starting with a ballet company. In our last meeting he introduced the idea of publishing some of my Ballet Class music on-line, for downloadable purchase. When I agreed, I didn't think that I would have to create a ballet company to go with the music.

I was adoment of not wanting to ever create a non-profit anything again. Jesus had to remind us that Mark has to carry forward some of Omhmar's missions. Continue the legacy that Omhmar began in addition to Mark's obligations to Griffin Cleft. Preserving places for Black People to perform the Classical Arts was important to him.

I agreed to consider the possibilities. The meeting ended with God prepping me for Jeremiah's return. He was as specific as deciding which bank and branch we should open our Corporate account with. What am I getting myself into? Everything is becoming so real. Accept the compensation.

As I ponder these requests, I feel a buzzing zap on the back, right of my waist. It seems to be zapping in agreement with me when I make a judgement or observation. Odd, what could be doing that? Continuing over a couple of hours, that zaps keep zingging me. I don't hear an answer when I ask. Then I hear God, say "Your turn."

Right afterwards ths suns shuts off. I mean, Black out! Curtain's closed, shows overd folks! Just for a second. Then it fades back in. I hadn't ever seen anything like it before. It really was if someone switched off a light switch, that's connected to the Sun.

After that and for the rest of the day. I kept hearing an announcer's voice over music narrating the end of a story. As if it were telling the end of my story. I kept interrupting him to say "That's not how my story ends." I recognized it from the movie *Big Fish*.

Later that night I go outside to smoke a cigerette, and escape the announcer jerk. As I am sitting in front of the Bodega, I feel Jeremiah's energy. "I tried to save you." I tell him. "I know." He replies. "Is Jesus with you?" I ask him. He doesn't say anything. We sit there a bit longer, before I say "I told God that you didn't have to die for me to get some help. We could have figured out another way."

"I didn't die, you did." Is what Jeremiah tells me. I run up the stairs in confusion. When I get inside, I don't feel that Jeremiah is with me anymore. And, I ask God how can I be dead? "It's impossible. You have eternal life." He replies, then continues. "There is a world where a Son lost a Mother, and a world where a Mother lost a son. But, no one has died. Only in appearance. Only in those worlds." Then I ask him, "Which world am I in?" "Neither."

Although, God and I communicate with more swagger in our conversations. He still speaks in parabells, riddles, and mysterious answers. I'm almost certain that it is because it has to be my choice, in most concerns. He's a pro when leading me on a thought provoking journey, that concludes with me deciding how he would expect, or want me to.

Even though, in almost every instance. I can say that it is my own choice in the end. Free wil, was definitely mine for the taking. However, we've reached a point where I've agreed to certain things, and the Spirit of Truth has assured me that it is too late to go back now.

That being said, Jeremiah is coming back soon. I'm excited to speak to him. Funny that we haven't spoken in years, yet I know everything that has been going on with him and his family, leading up to his transistion. That's why it is a blessing to have Two best friends. So when you fall out with one, the other keeps the Two at odds, connected.

Speaking of which, I need to check in on my West Coast family, to see how things are going with the funeral plans and such. I'm not big on attending funerals. I've already prepared my sisters, in the case that Mom goes beofre I do. I will take the lead in making arrangements for Uncle Gordon's transistion. And, I will not participate in any shape or form in Mom's. I will not attend her funeral, ever visit her grave, and have made it clear to

God. That I do not want Mom visiting or communicating with me from the otherside!

Not, really sure how that will work, anyway. Since Mom and I somehow have crossed over to the otherside together. Perhaps, she'll be like her biblical origin and live to be Hundreds of years old. And, I transistion before she does. I will continue to honor her, as I am commanded to do. I will continue the tradditions and customs that Omhmar has spoiled her with. But, both. Mark's Mother and Father is in Heaven.

When Jesus arrives. Each and everytime. The first thing that he wants to do is, plan my giving. They are constantly reminding me that God has chosen me not only to gain, and expand. But, to bless, and directly effect the lives of believers that need God's helping hands. However, I've been down this road before.

Already, Jesus has made me give more of my finances than I would care to give. However, he promises to give it back to me Ten fold, of what the Devil steals from me. Yet, That's a promise that I was clear on before. And, after giving Thousands of Dollars to the church, I can't see how it was given back to me. Therefore, I don't want to be that kind of Christian, this time around.

He and I have discussed alternatives to giving. I'm thrilled to donate sweat, and time to Non-profit organisations. Keeping my cash in my pocket. Or at least what's left of it. Last week, he and I got so upset at Mom, that I played Father Christmas in July.

Walking around the neighborhood reminding folks that God is Good, and giving them presents such as Clothing, Jewelry, gift cards, and cash in some cases. A dual task, that in my giving, I must learn who I can befriend. And, who must only know me. God says Lenders don't befriend borrowers. I clearly want to be a lender, and not a borrower. But, now is not the time for lending. Well none other than my time and effort.

I've chosen this year to Volunteer with The Bronx Defenders, The Bronx Republicans, The New York Phiharmonic, and The Universal Temple of The Arts. That's the after school program in Staten Island, were Diana and I took our first Tap Dance Lessons. The director presents the Staten Island

Jazz Festival every year. And, I've agreed to help her produce it this year.

I thought it would be a good way for me to familiarize myself with the Producing Hat. Unfortunately, the Presenter is impossible. If she is not having me show up for meetings that she flakes on. She's refusing to allow me to introduce her to the internet age of fundraising and arts production. This is an event that I will truly have to put in his hands.

Frustratingly, the voice over ending to my life continues. It's scary to think that my life is ending. But, where the announcer gets it wrong. Is that it is Omhmar's life who's ending. And, Mark now starts on the cutting edge of the New World. Now that I've had my psychic initiation, I hear everything.

God or Jesus will often interupt me responding, tell me that what I've just heard wasn't for me. But, when I mention that I am alone. Who else would they be clicking for in validation? They remind me that I am never alone. And, They also need God's validation for events, and ideas to come. As, They are working to prepare my future. And, they're not always guiding my present.

Hearing everything also means hearing negative, harsh comments that I wouldn't normally think of myself. My favorite place to exist is in pure positive energy. Having to overcome the challenges of the enemy's Two Cents into my knowing, are obstacles that I never thought that I would come up against.

All I can tell you is that, standing in your truth really does set you free. When doubt becomes the overtone of decisions. Knowing that you can't get it wrong and that you can't get it right, becomes the quentisiential divide. The dividing force between forward motion, and henderance. Everlsating life is the true reason you and I can never get it wrong. Concluding a judgement of wrong, means that you have ended a point in your day, chapter, or life. Where you can't make it right. Each day that we're given, is a day where making it right is a possibility.

God's process has a compact schedule where everyday of the week is assigned to something specific. During the Six month training period, Tuesdays were test days. If I did not design a test based on God's curriculum. They would provide me with real life tests, relating to lessons I've learned the previous week.

Now, that I have won the job of Griffin Cleft Managing Producer. We'll spend the next Six months working, and implementing the lessons learned to real life. And, now on Tuesdays, or as I call it Reward Day. I hear the masterplan unfold to include capable possibilites based on the effort I succeed in, weeks before. For example.

When pitching Hotel Nativity as a Film production this past Board Meeting. I had to make clear of, the meaningful purpose behind shifting all of God and Their momentum behind Seven years of beating the drum, Hotel Nativity The Musical. Seven Years of guided meditation through storyline, character anaylasis, touring schedule, script and music composition. Seven years with sounds that only hum on stage. Seven years of staging each moment in my mind. Over, and over again. And, Seven years of going away from my faith and love for Christ.

I've become so excited about the new scripts that I've written. Plus, a kick-ass, presenting process that is fresh, affordable, and fits right into the New World that God has created. Hotel Nativity lacked life, and was collecting dust on a shelf somewhere.

You and I can see why, They wouldn't just say yes. There are those in the Christ Family, that love the show exactly they way it is. And, would vote, and urge for the consistancy that is the original plan. However, as I stand in my truth. They all can see that it is either Film, or toss the shit in the trash.

There's no way that it can go on my theatrical radar. I'd toy with the idea of opting it for a Broadway or Off-Broadway run, produced by someone other than Griffin Cleft. But, then we'd have to change it so much, that it would play like any other Christmas re telling of the Birth of Christ.

However, my depiction of the Immaculate Conception is that it transcends an Annual celebration. And, becomes moreover a mantra for daily living. Abraham says, that the resources for entering imgages of negativity into your mental hardrive, are abundant. But, if we could worry less about what other people are doing, and instead focus upon our singular purposes in the time expanding universe. We'd be better off.

So, the show takes place in beautiful city, seeming to sit about clouds. Called, Nativty City. In it, exits the Hotel Nativity, which in the opening number is left to Jewelius Carpenter II, after his father is murdered.

By intermission, I have Mary dying in childbirth. Trying to save face, for plans going south. Mayor Maxine Diva awards the child Immaculate Conception. Nativity City, a town themed, commissioned, and instituionalized after the Christian Nativity. Have on their books a law called the Immaculate Conception Rule.

Where there could be scandal, this rule overides any question. Sanctioned by God, whatever the truth behind the conception, Immaculate Rules. The Second Act fast forwards Eighteen years later to where the child becomes an adult. Then follows, Jewelius, Mayor Diva, and JC III through out their missions to keep a city in the Black. Make a Family's wronged leagcy right. And, regain the charity in which Nativity City was founded upon.

However, all day today. I have been getting visions of cascading, sweeps, through out wide landscapes, that fly upward and out after Mary dies. I'm seeing a new opening, that plays the overture with the opening credits, that fade in and out of scenes, where Jewelius is crashing his car. Plots that are certainly limited by any staging of the Musical. Limits that only recorded tape can break. I can see that the answer is yes. The Board agrees. I wonder if God always new that each of the Six times that I wrote Hotel Nativity. That we'd be adapting it one last time. To film.

On Wednesday, I am led by Abraham. Through the teachings of Esther Hicks. I spend the day channeling mostly pure positive energy. My favorite kind. Rolling us into Thursdays. Griffin Cleft Day. In our presenting process, we've chosen Thursdays to be the start of our work week.

Since, our stage tours will open in new cities typically on Thursday evenings. We follow suit with episodes on our Youtube page, Open Isle Channel. Premiering Thursday mornings. Along with new press and promotional campaign kick offs. I told you. They really did think of everything.

Also, on Thursdays, I wear the hat of Managing Producer, executive for the bulk of the day. Or, at least between the hours of Nine to Five. Where my physical staff slacks, my spiritual staff pick up the pieces. Although I can not see them. I hear them. I feel them. Depending on the breakdown, I can really feel the energy of wall to wall crowdedness, of a busy, New York, production office.

I picture the set of *Mad Men*. Good show. God is of course, Bertram Cooper. Jesus is Roger Sterling. And I am Don Draper. I guess Jeremiah, will now be Harry Crane, I suppose. But, that paints the perfect series of images, relating to how I process what is happening. I say that I am Don Draper, not only because he's the sexiest, and hung! But, also because he's always expected to come save the day. Again, and again, like it's nothing.

They keep telling me not to care. I must admit that not caring usually promotes for more productive, and easy breezy days. But, passion is what gets me out of bed in the morning. I love my life, in the way that I have always enjoyed the luxury of going back to sleep, if motivation doesn't meet me at the point of bright eyes and bushy tails. It's a great trade off that makes being poor and bored, almost seem worth it.

I kid, even when I performed Eight shows a week, I still didn't roll out of bed before Ten, or Eleven in the morning. However, the procrastination that leads me to a Three day window of no sleep at all. Makes me the quick study, and why Jeremiah and I would call ourselves, Last Minute Divas.

Jeremiah liked to say that, "It's going be, what it's going be." His mantra for excusing our last minute preparations for tasks that should be given staff and time. Like, Griffin Cleft's mission clearly states. The Curtain Must Always Go Up!

Fridays, are my time to spend with Jesus. We spend a great portion of Fridays eating fish and studying. I hope that I am catching his guidance clearly. I've already mistaken his study topic of Sumeria, with Samaria. Thinking that because of the name Omhmar, my roots date back to Muslim, and other Abrahamic faiths. And, of course making the mistake of studying the Devil for Six weeks, when God only wanted me to know that whatever has happened. Has in fact, happened.

However, that study. Which led me deeper into our past existances, more than I would have wanted. I did find eveidence of something kind of helpful to understanding how this thing with Jeremiah is supposed to work. They say that the Devil was the soulmmate to King James, who designed the version of the Doctrine that my family applies. And, as I heard the information. I was given the same visual of an Earth where one man walks, and the other is shadowed from the mans feet. And, you can clearly see that their steps are together. However, one image that you see fully manifested. While the other you see only part of what is there.

Which makes more since to me than God making Jeremiah the Left Side of my body. Knowing that he can do all. I wouldn't deny my belief, if that is it to believe. But, the companionship that I got just from the short time that we've spent together, since he left the world. Gave comfort to my lonliness. Will I feel his absence if he goes into my body, while discarding half of who he is?

Sunday's and Monday's I get an extra boost of God's care. With the Mother on Sondays, and the Father on Mondays. And, alone. Without distraction, complication, or assitance from Them. And, now on this Monday. God and I are almost alone, accept for he has returned Jeremiah.

Here at Griffin Cleft NYC Headquarters. Our Earth, and Christ families are standing by witnessing God's wonderous Glory, as he brings to fruition. The eventful idea, that Omhmar+Jeremiah=Mark. As though God, just plopped into my living room.

I feel the energy that is Jeremiah enter the building. Forgetting all about Mark. I completely become Omhmar. And, Mongo Jerry, and Rumpleteezer are back at again. More excited of how fierce the magic of Gods design is. I totally forget, that Jeremiah is not really in the building. Or is he?

When I ask anyone. Including Jeremiah, as to what his world looks like. And, where he is. I only get the reply, that he is standing right next to me. He see's what I see. Feels what I feel. However understands more than I do. And, wishes to take the lead. Here is where we first bump heads.

I know that God must have been training Jeremiah before he transitioned. Juat as he did, I. However, our communication is not so much practiced, that I truly trust what I pick up from his speaking. Yet if I think, wow. This guy that I am watching on the telvision looks oddly familiar. I'll have a vision float by of a trick, that Jeremiah and I picked up at a club once. And, then I would say. "You're right. He does look like that guy. What was his name.?" Then I say, "My God, I forgot his nameith guidance I begin to think, it may have been Jordan.

I would anticipate that the action must be communication with Jeremiah. He brought me to a single event that only he, and I experienced. And, it's the first time I've thought about Jordan since hardly meeting him those years back. Now, I must take a second to tell you about some physical feelings that occur when Jeremiah is around.

Firstly, those zaps that vibrate the left side of my lower back, still occur. I should say that if right is right, than the left should signal a no.

Then god reminds me, that in my aswering. I should never hear or indicate the acceptance of a no. And, he did say that Jeremiah would occupy the left. So, when I get zapped on the right, I say God thinks so. And, on the left, Jeremiah must think so.

Where busy kee kee-ing the morning away. When their arrival actually interupted my morning routine. I better get back to it. It's funny how in my reality, our closest friends use to have the problem of needing to separate Omhmar and Jeremiah so that they can be productive.

I remember at an awards event that we'd attend every year at the Los Angeles Music Center. Jeremiah and I hadn't seen each other for a while, do to our travel schedules. And, as Mrs. Four Octives observed our disinterest in the, very well off guests in attendance, she tapped Jeremiah on his shoulder. "Baby, could you reach across the table and grab me another biscuit? Thank you honey." Then she turned to me and grabbed me by my Neck tie, and said. "Do you really think now is the time to be catching up with Jeremiah, when you are sitting right next to the Vice President of Bank of America?"

Yep! Just like that, our souls continue to tangle in most inopportune moments. God gets us to inccorporating Griffin Cleft. Of course, any legal documentation that I had for myself, and the company. Is now void in the new world and we must do it all over again. That's why God keeps telling me, that if I really want something. I have to often try succeeding at it twice. And, probably after the third time, walk away from it and try agin, when fresh. At another time.

To my understanding I think that as I am witnessing Jeremiah's responsibility in our New World. He has to play the role of people that don't yet exist. For example, when I am speaking to the young lady on the other end of the phone, helping me to document Griffin Cleft. I think that it is Jeremiah on the other end. I'm not sure if his purpose for leaving and returning throughout the half hour process that it's taking, with lots of elevator music holding my ear. God, interupts us to tell me, "Everytime that you ask Jeremiah something and he has to return to answer you, you're distracting him from doing his job that requires him on the other end. So for the time being of this phone call, you should try to rember that he is not here. But, can hear you. But, hasn't had enough time with you to know when to answer your call, and when to let you figure it out."

As I take a long hot shower. I imagine Jeremiah flying in and out of my windows that I left cracked purposely. And flying over the country to be in LA, or New York, or Mexico. When our business on the phone is done. Jeremiah hops in the shower, and it feels as if we are taking a shower together. That, weirds me out to the fullest. I hop out of their as fast as you can say Holy Ghost!

What the fuck yo?

You know what I was thinking? That on three separate occasions, I attempted to committ suicide. However, I think the real reason that I failed, besides fearing the unkonw. Was that everburning question of whether God will forgive us for self murder. It dawns on me that if I died from that bullet, that was shot from my own trigger.

I actually killed myself. With that, I guess the saints can relax. I think he would forgive us for it all. It's the reason his son died on the Cross. Which is an event you may want to not mention, if you ever hang out with Jesus. I get the feeling that he's still not over that!

Jeremiah and my first task, seemingly went off with out a hitch. But, the evidence remains to be seen, until the arrival of our corporate ducments from the State's Capitol. Until then, I think that God and Jesus are pleased with us so far. Throughout the day, I carry on with business as usual. God picks me up from time to time, to go on an adventure with he and Jeremiah.

Sometimes short walks around the block, pointing out signifigance in the sky. There are days where I swear that I can see where one world ends, and another one starts. Separate clouds varying in color and shading. Many brilliant, and odd occurances happen with seemless timing to our journey. Occurances that constantly leave me wondering. Who is really helping God?

Twelve white sedans. Making a right at the stop sign, accept for every Fifth one. Speaking to me in terms of colors represented, and numbers. Other things too. That somewhat culminates, my independent studies with God. God has given me an anallagy to share with you. As he did with Jeremiah and I on this Monday.

Imagine that you had the ability to control animals just by imagining what you want them to do. You close your eyes, imagine that a small bird landing on the red car parked in front of you lands, and chirps at you. Without opening your eyes you hear the chirp of a small bird. Then as you open your eyes witnessing that the bird has obeyed, the bird fly's away from you.

The odds that on a sunny afternoon, standing next to a red car while a small bird stops on it to ask you for some food are actually higher than you think. So if the bird was under control to inact your specific thought, the success of it's obedience would go unappreciated. Simply because no one expects that a bird would do much more than fly, land, chirp or pirch.

Now, if you closed your eyes and used your powers to control another animal to do the same analagy. Let's say, a Human. You would open your eyes to find that some distraction has interupted the thought track, that yielded the Human to obey. The Human feels the energy of it being summoned. However, the energy is often much weaker than the energy of infinite Humanly actions, that we prioritize over the action of connecting into our inner beings for specific guidance.

Well, God is a busy man! And, can't conduct my every moment. Moment to moment. And, in real time. What he can do however, is provide us with Soulmmates that force us to connect, and yield to the asks of our inner being. And, spirituality that is the other Ninty Percent of our brains that, most bodies never get use of.

Jeremiah's decisive, and clear manipulation, leads me through conversations, actions and feelings, that make me wiser. Better. Stronger, and much more dynamic. I think that there is a physical additive to his power as well. I took a balance today in a Pique Arabesque. It's a ballet position that any professional would be able to hold for at least Twenty seconds.

But, as minutes upon minutes passed, and I were still in my beautiful Arabesque. I wonderd if the science for what makes this possible has been unblocked to me. Or, if Jeremiah himself were standing there. Supporting me as I stood on my tippy toe. Whatever the case. Bitch! We bout to be Extraordinary!

Being that it is now, Tuesday/Reward day. I can here the achievements being praised of Jeremiah and I successes in understanding, communicating with, and working together to provide the forward motion in our process, that has been severely missed. After work today, we kick back like old times. Turning the hallway into a cat walk for which we stomp. I can feel him battling me on the runway, as Pandora bumps our last, mutual soundtracks in the background. I suspect because of his doing.

Jeremiah has expressed that the job title he would hold, is Griffin Cleft's CTO.

Out of corporate structure for so long, I forget, and am unfamiliar with a lot of the new positions created by our technology age. Jeremiah is not. He knows the ins and outs of multi-faceted internet and mobile technology. As well as statistical data, privey to the finest of Ad Men. A hard pill for me to swallow, since in our previous pairing, Omhmar was the smart one. While Jeremiah just looked pretty.

Jeremiah accessing the same source of infinite wisdom that Abraham exists in. He has access to all the knowledge, that ever was. The way he can guide me in, and around a mean search engine, to find exactly what I am lacking. And, so very quickly, is incredible! I'm certainly impressed, and excited about the possibilities. Yet, some how, I still can not shake my perception of Jeremiah, the best friend. The one, whom only took guidance from me. Amd participated very little in our decisions.

Now accepting the fact Diva knows infinitely more than I do. It gives me question to think of, who was really chosen afterall? Jeremiah has been giving me missing links to our story. He's told me that this was never God's idea.

In our beginnings on the other side, we were friends. We along with a great deal many of other folks came up with this plan. But, spearheaded by Omhmar, Jeremiah, and Mark. We would enter Earth at similar times, and be provided with a similar upbringing. But, only on different Coasts of the United States of America.

Once we became of age, we would attract ourselves to find one another through our Souls searching for one another. With everyone on the otherside paying close attention that our dischord only shortens as our natural, and deliberate choices lead our spirits to join one another.

Suceeding in this event would promote the evidence that our souls could do it once again, but this time from different worlds. Instead of, different coasts. Again, telling the Story of Jesus Christ. There lives long commitment, Mark and Jesus are at it again, to expand on Human origin, evolution, and God's pupose for creating the World.

So, as we pitched this plan to God from the Source of who we are. We were approved, and chosen which began this Journey, Thirty-Something years ago. At Thirty Two, we begin another life cycle as a Man. This is based upon Christians basing the origin of Man upon the Man who was Jesus Christ.

Now after completing my Saturn's return, and Thirty Two year Cycle.

Jesus, God, and everyone has decided that the work that the Two of has done on Earth, has become substantial enough to begin our work as different, beings. Beings, that between the Three of us and God, gives Mark that power to access much more than the Ten Percent that our able bodied redily limits us to tap into. With the added power, we can evolve much faster than the slow progression and expansion of our last Two Thousand years.

Now our work truly begins. It is quickly becoming work that is absolutely logical to the goals that I seek. But, never in my wildest dreams, did I ever imagine that I would gain new, and strong passion for new work and corporate missions. The projections for Griffin Cleft are out of this world! And, that is only the start of what God plans to use me for.

He assures me that if I were to be come America's first openly Gay President. I am currently at the same place that Obama was in, when starting his journey at the point that God's timming would allow for America to ease into the idea of a Black Man running the country.

Don't get too alarmed. I am so grateful to those who accept the challenging, underpaid job of President. I however, have never had that dream. Not even once among the many dreams that I have, and have had.

As Jeremiah and the gang, continue to beat the perception that I am the one who has died, and he lives. I am starting to come around to idea, which some evidence now shifts my focus to. Small events like, today I get a signature card from the U. S. Postal service that is to Omhmar, from Omhmar. Signed confirmation that a package was delivered, and an Omhmar signed for such a package. However, I never sent or signed this signature card.

I got a letter from California asking me to collect a payment that needs to be settled into my estate. Like money coming to my family after my death.

And, finally, a small detail. But, an important detail. When I enter my building, there is a long beeping noise after I swipe my key magnet. It alerts us that the door has been unlocked, and can enter the building. But, it's a long high pith beep, that echoes the exact sound that heart monitor machine makes when a person flatlines in a hospital.

Maybe nothing much to you. But, my senses are in over drive. Every smell, sound, image, and taste, sends me to the thinking that I am being communicated with somehow. Whether it, be a smell that reminds me of my Grand Mother's house as a child. Or, the smell of an apartment burning down, that I just can't get to. But, still doesn't stop me from running around my building. Knocking on doors, while yelling, fire!

The severe smell of a fire nearby, yet every apartment in the building has been checked. And, only then does the smell disapate. An event so dramatically played out when they were simply trying to say, you're stressing. Why don't you take a break and smoke a cigarette.

As minutes become hours, and then days that pass along our intimate evolution. I begin to allow moments where I submit to Jeremiah's leadership. He couldn't really hold a tune, too tough, before. But, now Jeremiah's become the expert voice instructor. Who gives me those finishing touches on my tone, sound, and pitch that I have been working for, for years.

If young boys who are like we were, had fathers for guidance, or financed our guidance. We would get "It" right, within the first series of trying. But, through out my life I've questioned observations of Television, or social scenes, and wondered very specifically. And, then tried immulating those observations I'd perceived, in life. Yet, only to say to myself….. "That's not quite it."

Now if I want to sound like fiero in Wicked, Jeremiah brings someone from the other side, that has done Wicked to coach my exact pitch, tone, and inflictions. Leaving me to now often say to myelf, "Oh, that's what it takes to sound like that."

Since, I agreed not to study Colors during the Purple period. Jeremiah, can now complete the Colored spectrum of my Griffin Cleft's design. I did not know which accent colors I would use. But, I knew yellow was to be a part of his design, because the only yellow State on my Map of the United States is New Mexico.

When God first began the Purple Period he told me that Jeremiah and I would meet angain in Mexico City. Before Jeremiah's transistion, he toyed with the idea of being creammated. And, thought that his ashes should be scattered in part of Mexico City.

He chooses an almost Royal Blue, just a tad lighter, to go behind the music area, of the Studio. And, we continued the same yellow/gold foil Faux finish, from the bathroom, on the Rehearsal space walls, and on the North Wall, we have Three Chalk Boards, cut into The Faux from the living area. So each way you turn in Jeremiah and Omhmar's room. The energy calls you to different productivity.

What I am calling Omhmar and Jeremiah's room is offically called the Four Octive Family Rehearsal Hall, during the AM. And, The Kennedy Room, during the PM. I originally named it Kennedy after Jackie, O! But, when another great friend of mine, Tracy Kennedy transistioned only weeks after Jeremiah did. I knew that it must have been foresight, knowing that Tracy would want to have me teach this process to Mrs. Four Octives. Tracy's best friend. Kind of the Sixties, "Black Panther" versions of Omhmar and Jeremiah. Tracy and Mrs. Four Octives mixed business and pleasure just as well, also.

I've been clear, that my continued interest in learing, or education, has been through the method of Autodiadects. Spiritual guidance that carefully directs you through the lessons, and origins of what I must learn. Including Course Study academeia. And, now we can add vocal coach to the programs offered in the School of Life. I feel music inside of me, in a way that I've never felt in my life. And, all for Free, Ninty Nine. The most affordable rate in the market!

Jeremiah goes off the deep end sometimes, as he shakes and stumps, and throws Diva tantrums. He keeps trying to tell me that there is someone else here. Whom interferes, and Because I don't pick up on the extra participation. I blame Jeremiah for someone else's doing. I'm not sure what Jeremiah's job fully intells, but I am sure getting people, in which I can not see. Out of my house!

I know it's not the devil. So, who on Saturn could make Jeremiah so fearful and doubtful. He's expressed that he's not on the Jesus tip anymore. I can understand why not.

Yet, we're both supposed to be working for God. Omhmar + Jeremiah = Mark. That's what I was told. "Exactly! Not Plus all these othere mother fuckers." Jeremiah interrupts.

As the week continues, I wonder who could be the person Jeremiah refers to. I haven't felt Marcus since I asked God to take him home. Could it be AC? According to Facebook, AC made it back to his Mother's house as well. In this week's Board meeting I bring it up. God decides, that none of us can do more than our part. If Jeremiah has a problem with someone God has chosen, he has the same outlets for difussing the tention, as I do.

I've heard them say it before, but didn't think that it was for me until I heard Jeremiah say it. "They hang around you because they think that you are Jesus." He tells me. "What? I'm Mark! I report the story of Jesus." As I say that, I think of Elton John's *He Lives In You*. Or, Stanley Bennet Clay's, *Looking For God*. A stage musical he wrote in the Seventies where the lead character sings the assimilation that he is God.

I'm so happy that Jeremiah is here to now help me put the commercial viability behind some of my dynamic ideas. Jeremiah was and will forever always be a pop culture Bueen! He studied commercial branding within Showbusiness, better than Joan Rivers, herself. Whenever he had to wait a moment, he'd either stretch or read a magazine. He could name off brand designers from overseas, and even call to evidence the brand's runway and print Models. He knew salaries, and networths for all the movers who roar.

He had very particular affirmations that kept him trendy. Like, "I only keep beat pieces in my closet. That way, whenever rushing, or approaching laundry day, whatever I put on will make a beat outfit." Adoment of taking his clothes off the minute he got home, so that he wouldn't stretch out his shoes, or dirty up pants that can be warn again, if kept pressed and wrinkle free.

Glamorous and prissy, Bueen was the most. And, since he has arrived, he's mostly been reading me for filth.

"I'm in Hell! Jeremiah says. "No! No! Damn, it does exist? And, faggots are going with gasoline jocks?" I panic in reply. "No, stupid! I mean forced to stay inside, doing a bunch of nothing everyday, is Hell." Jeremiah continues, "What happened to the Omhmar that planned every waking moment of our lives, when we filmed Working Birls?" I move quietly off to the Kitchen, and up towards God…. "Why does he think that I am Omhmar?"

Between the saints thinking I'm Jesus, and Jeremiah thinking that I'm Omhmar, I need a day off. "That's a good point that I was going to bring up on Thursday, but since you mention it." Jesus says to me.

Jeremiah will need a day off. So each week, he will leave on Thursdays, and return with Elijah on Saturday mornings at Twelve AM." God Says to us. Of course, so that I can have less distractions at the start of Griffin Cleft's week. Wednesday night at midnight, I walk Jeremiah out.

Since becoming chosen, I've adopted a ritual of being the first to say the Lords Prayer, right as the clock strikes midnight, each morning. We joke, that the day doesn't begin at Nine, rather Twelve. I recite the prayer outside, as I look at the sky, and neighborhood. Looking into every adjacent flickering light, as I wish them all blessings and good fortune.

I wait Seven minutes, to pass before going back into my building. This allows for God to reset the day, and transisition all that is summoned into the absence of all that was. He and I take turns in a way, because I prep the home each Saturday for his return and rest. My Grandmother use to say, "The door's open. Come on in Jesus." Well I think that the Christ family has decided that Griffin Cleft-NYC, is the base for Operation World Peace.

I'm volunteering today at the New York Philharmonic. The season opens this Month, although I wont be back to see a performance until December,

The Purple Book

for Handel's Messiah. As I place roses and thank you cards on the seats of sponsors and subscribers. I begin to think about this weeks Board Meeting.

When I had Lambco, we use to hold an event called The Habitat Demonstration, once each season. It was a free studio performance open to the public where the audience would get to see new Ballets as they developed. As well as, a Lecture Demo on Classical Ballet, The Profession.

Before Jeremiah arrived, I started doing them before each Board Meeting, and filming the demonstrations for Mad Science. God wants Jeremiah and I now to plan, an artistic vinguette relating to the Artistic intent of our timeline.

Since, they've sprung this Ballet Company on me, I've decided to choreograph a piece of my most favorite music written. The Italian Concerto by J. S. Bach. We begin with only the Second movement this week. For the narrative, I am choreographing an invention that tells the story of how the Earth came to exist.

Still out of practice for most everything I need to be an aficionado in. I venture forward. If God says the time is now, I'll trust him. However, how cruel to force me to dance on camera, for the world to see. When I haven't done as much as a plie since 2010. My neck and my back ache. I want a Hundred and Fifty Thousand! What's worse is that, the story evolves so quickly and fully, that I don't remember the bulk of it.

Thankfully, J. S. Bach did his part. The Second Movement of BMV 971, moves the spirit, through the soul, and out the finger tips. Energy that slows, but never stops. Melody lines that gradually step down towards the Top. Modulation after modulation. I enjoy listening to the Baroque period most of all, because the math makes since. Once we started copyrighting intellectual property, Composers said fuck the math. And, others like Stravinsky, said fuck the melody too.

But, Bach guides us safely to the start of another Board Meeting. Again, it feels as though the room is full. I use sounds, visions, math, and electronics to include their opinions and votes. Yet, Today there are Three Vans beneath the Four Octive Family Rehearsal Hall. Blue and White, Red and

White, and lastly all White. The Vans immediately put me in the mind of my Stepfather.

He was one of those cool Vets, that spent most of his day, cruising in his van that looked like a ride at Lunar Park! Lights and speakers underneath. Whenever we as kids got to ride in his van, I'd feel like we'd gotten a First Class reward. So although, King was not in the driver seat of either of the vans. I was picking up what they were laying down.

The person that Jeremiah doesn't like is Diana's father, who passed away in 2012. I had no idea. But, from his funeral service then, until now, King has stuck with me. That's what we call him. So, the big issue in today's meeting, is King sticking around. Jeremiah, feels that if King is going to be apart of what we're creating, he'd rather not work with me.

King just wants to help, and Jeremiah is a stickler for the rules. Omhmar + Jeremiah = Mark. Not King. But, here's the deal. King and I have been through a major transformation. He has lead my teachings miraculously. He's encouraged me to stay positive and really excel at living the perfect life within the Law of Attraction. He brought me back to the church. He saved me from waiting for the Big Break, and back to creating it for myself.

King, also got Secret Societies to pay attention to me. When, I speak with king. It often feels as though I am speaking to God. He has only my best interest at hand. And, he decides that on how I rate my interests. I can't see a problem with him, or the need to ask him to leave. Meanwhile, Jeremiah is convinced that an altermatum is the only way for him to stay. Since we're also discussing setting up Jeremiah's estate, and financial terms of the process. Knowing Jeremiah's limits to working and helping, is paramount. I am not going to give any percentage of my hard earned money to him nor his family, just because we were friends.

That's the type of Jesus shit that will leave your ass broke everytime. He's going to have to take the resident Left Side, that God has allocated him from my body, and be the soulmmate to Omhmar, that Guides Mark through God's Process for Creation. Right now, he says he doesn't want a board seat. He managed to close out pre-pay credit cards that I opened specifically to save money for his Mom. Who hasn't gotten back to me yet.

In fact, I really haven't spoken to anyone in LA to confirm that Jeremiah really died.

Since he thinks I am the one who has, I keep hoping that God will take me back to a World, where Jeremiah stands statuesque with wide eyes. And, this experiment can end. I don't know how to diffuse beef between spirits. But, I know that my CEO better do something. He promised that every next step would seem logical. And, this seems highly irregular.

I take a break from the meeting to call MaWest. I brought over some documents from when I first built a management company for them, which had a cell phone number on it. She actually picks up. I can't believe that it took more than a month just for me to try that.

Aside from giving her time and space to grieve. I wanted to offer her some blessings, as well as know what the situation in Los Angeles looked like for her. When I told her of giving the Management Company another shot, she seemed excited. However, she mentioned possibly not being able to trust me. She keeps it real! That's understandable. I gave all of the info to her assistant, who gave me a yes to move forward. I then signed Two young boys, and one young girl to get her started.

I sent over their contracts while CC'ng MaWest. When she got the e-mails of the contracts, she quickly sent me an e-mail to ask me not to use Jeremiah's name in anything without talking to her Lawyer first. That's not the hardest ask for me. If you have got an attorney on retainer to estate plan and incorporate for Jeremiah, than he can do it for Omhmar and Mark as well. I welcome the opportunity to speak with him.

Week after week go by, and no call from a Lawyer. I say screw it!. I send the Three kids their drop letters, and fille the company's inccorparation back where it was collecting dust. As I move forward with the rest of the commodities that we share, I'll have to see how we progress. Of course, Jeremiah is in the back ground saying Fuck that! Do what you wanna do, you can use my name. Infact, use my middle name too, Kyle. This doesn't sit well for me. So I begin to question how we move forward without MaWest consenting?

By the end of the Board Meeting, Jesus gave Jeremiah the task of getting his mom to come around. I think this is where we begin to move to quickly. As the Two of them begin to understand the physical event that has taken place. The communication is very lacking between Jeremiah and Omhmar. They feel as though they are really speaking to One another. However, they really are only basking in one another's joyus familiarity to their relationship.

As I try to navigate them through this. The task of manipulating someone as strong willed and doubtful as Jeremiah's Mother, into jumping on board a process, idea, and plan that has yet to take shape or it's legs. Really is not a fair ask of Jeremiah, nor anyone else. However, Mark, still on Earth and connected to our Human interactions, simply can not move forward without respecting MaWest's requests.

As God often tells us, if you've driven them to actually file a suit, it means that you are worth something to sue for! But, still. Call it the Jesus piece in me that has me to honor my Mother and Father. So that my days may be longer. Has me kind of by chains. Since there is still so much to be decided upon. I decide to go ahead and cancel filimg of Mad Science.

Partly, because I didn't know that Jeremiah would die, when I began filming. Now that he's here, and in the mix, I need to see where and how he fits into the groove that Jesus and I already have going on here. Also, because as I watch the credits roll up to read that God is given title of Director. I would want a better product to represent my King. Can't let you know how right down, and rachet my Father can be. Hapenstance Central!

Things seem to be settling down between King and Jeremiah. Now that the cat is out of the bag, and Mark, don't give a damn. Perhaps, the others can take a clue from us. Jeremiah's new attitude with me is now ofcourse because I have canceled Mad Science. The show that discovers it all. He's ready to jump on in head first, and produce the same tired shit that gave us such a standard company before.

Cheap technology, and user friendly interfaces promote more resources for being spectacularly, brilliant. Evidence of us living in our Grand Design! I want people to look at the Griffin Cleft Brand, and really say, "Them folks over there get the curtain up, everytime. That must be God!"

I'm certain that he would want that too. Jeremiah just wants to fill all of my days with answers to questions I wondered my whole life about. He gets so mad at me when he presents me with the answer to a question that was burning in my heart a week ago, and I dull his Ah Ha, moment with . "Bueen, aint nobody got time for that!"

I'm always on, what the fuck is next. While my mind circles forward and backwards, faster than what's next. So I realize, one could say that I have the problem of asking, moving on, and then forgetting that I asked when my vortex fills with the very thigs that I have summoned, and asked.

Funny how I just realized that I have mastered another ask, that I had a couple of years ago. How do I live in the moment? Most of my moments have my mind thinking about the future, and my energy stuck in yesterday. But, moving on the way we do in God's process, certainly has me living in the moment.

At this very moment I am excited, because Omhmar, Jeremiah and Jesus have decided to reveal who they have chosen for me to Marry. Thanks to Facebook, Jeremiah is able to lead me to a face from my past that I forgot was even a FB friend. Also, making perfectly good sense, since my original full name was John Mark.

A young man named John, in Los Angeles was once a Fourteen year old, heart throb. A hunk that had it all. Looks, personality, drive, sex appeal, and very charming. All at Fourteen years of age. I wanted to sink my teeth into him upon first sight. However I waited. Year after year. Prom after prom, to finally graduation day. Not high school, but College. Ripe and ready, I fanally plucked that sweet Johnathan!

When we cuddled in ecstacy, afterwards. I told him that I would marry him one day. A memory I only regained after talking to him. John reminded me that I said that I would marry him one day. Sweet! That bit made an impression on him. How perfect, that he has settled into the idea that he could be married to me. At this time, Mark still thinks that we're doing some kind of Hollywood, Illuminatti-type, fake arrangement for personified life.

So, no doubt, my conversation with Johnathan was all about how we wouldn't have to be monogamously romantic to one another. His concern would be more red carpet optics, and such. But, that is not the conversation that God was seeking to invent in Mark's new world. In fact, he told Mark to stay out of it.

But, still losing more face each day, in the miricale that he's been blessed with. Mark fucks it up. Although, lil Johnny agrees. God and the family has lost interest because I chickened out! Afraid to stand in my truth and see if the boy could love me back. I give him an out, to not love me or present me with other rejections.

Of course I'm fine either way, because my sexual proclivities have surpassed the vanilla interest of young John. If God or anyone else thinks that I am going to try to make a monogamous go of choosing just one? They're wrong!

You know, that he will have to be a bonifide freak, to match all the freak I got ripping through my loins like old Stink Face, ripping through some Popeys's Louisianna Fried Chicken! With the addition of the new beings attached to me, they increase my freak! Eventhough, God claims this aint about sex. I digress.

Business continues like this, with lots of dram, arguments, and negativity thrown into the mix. However, as we round out those last couple of months of the Third Quarter, we have a great deal of work to do, doing this pseudo vacation time. I'm often confused as to whether we are moving too fast, or too slow. If my friends are any indicator as to how far we lag behind, we're moving at a snail's pace. A boy that I gave his first job to in Los Angeles, has just won a Tony Award. Mrs. Four Octives has relocated here family and company to the Bay Area of California, and aint missed a beat. Through the recession, and his Mercy, she now has more funding. Stronger Board presence and sliding her ass right over from the Executive seat to the Artist's podium!

I know that I am not supposed to judge my life according to the lives of others. But, as I can't turn on the TV without seeing another close friend making it big, I wonder if God intended for me to be in that group. I mean the entire circle of us struggiling artists in Los Angeles have now been cast in major projects on HBO, Showtime, Own, and Principle lead work on Broadway.

God needed that time spand of years for Jeremiah and Omhmar to sever ties. First of all, if Omhmar had been by Jeremiah's side. They'd still be walking on Earth together. Keeping Jeremiah safe was an easy ability of Omhmar's. But, in the off chance that Jeremiah was to go while he and Omhmar were best friends. Omhmar would have thought that the energy channeled and called Jeremiah was familiar recollecting while mourning such a lost.

This way, it proves to Mark, and everyone else that what is really happening is a real live event. Because, Omhmar and Jeremiah can hardly stand one another as things go. Jeremiah is constantly reading a bitch, and stating authority. While, Omhmar has checked out and lets Mark mill about mostly unguided. And, Mark, poor baby, is trying to maintain his positive, delightful disposition while the Christians work his nerve. But, all together the Trio work steadfast in their carefull, slowly unfolding launch of Griffin Cleft.

Chapter 12: You Win Some

Damn! I'm appauled at how the prices have risen for production equipment that I purchased years ago. It hit me that Jeremiah probably doesn't have my equipment stored in his garage anymore. Now the inflation has me thinkng that buying twice feels more like buying Three times. There has to be some advancement in technology that will cut my costs, while efficiently producing music, at this practice level.

I'd almost forget about God's promise as Mom and I walked away from Guitar Center. Shamefull of these immigrant Taxi Drivers, to only exercise refusal to us Blacks. I almost wasted God's ear complaining about it once again, when. I look behind me, and right in front of the Trader Joes, on

Atlantic Ave., I get salute from Louie 234. My sexy, future ex-husband/in law/babydaddy/co-parent, that is visiting Brooklyn. Yet, he lives in an even more ridiculous city, to be a Black Man hailing a cab.

Louie is some one that I had not spoken to in Two years. I left my phone in a taxi around that time. Loosing many great contacts of Omhmar's. Add to that. Louie 234 is someone that I really hated loosing. I've been trying to find a way to get a hold of him, this entire Two years that we've been a part. Unfortunately, he's not here for vacation or celbration. But, instead he tells me the solem news of his Mother's Transition.

I arrived so early to meet Mom in Brooklyn, since I tell her that it's the halfway point between us and Temmbucktoo. It's a lie, but. It does beat getting on some antiquated, sluggish, Walrus-float. That trails so far behind the fast, paced, cosmo needs, it will make you hate the Statue of Liberty. It's lovely, if you like that sort of thing.

I almost prefer oppression. As I sit on a gorgeous Brownstone Stoop for nearly Three hours, I wait to take my Mother to lunch. Each time my comtemplation of ditching her, to return to my elegant barrio, concludes with me looking for a train station. God would pop in to my ear to say, "Who's going to judge you if you leave? You know that she won't. But, you will miss the surprise that I have planned for you." As usual, he seemed particularly, amped about his good taste in gifts. And, I didn't want to offend Two Birds, with One Brownstone.

So, here we are with another Momma's boy, becoming Motherless. Yet, he's around our "Gay Daddy" age. Yet, that doesn't prevent him from turning into a little Bueen, whenevr he would share his colorful stories about his Mom. You know how some people can describe a family member so vividly, that you get a since of who they've described?

Whenever Louie234 tells me stories about his younger years in New York City. He appears to come from a progressive, Upper Class, Brooklyn Black Community. A family that supported and, prepared Louie234 for a fabulous young adulthood. Dippin it, Doing it, for Iconic Fashion Houses. As well as, iconic celebrity artists, such as Annie Lebovitz.

In High School, Louie234 studied Aviation. He's tall, built like a Tyler Perry Production lead. He's a tall drink of hot chocolate with education, sophistication, and class. He definitely has the swagger of a King. Once upon a short time, I thought that he'd be the one. But, I fuck that up! And, now it's so sad to see him in mourning.

He returns to D.C without coming over or, us seeing each other again. However, he didn't leave before we could exchange numbers. And, he was the evidence that God can put you in the right place, at exactly the right time. Street encounters are hot. But, right accuracy is power.

This morning I am moving a little sluggish. It's my Birthday, but not the fun ones. I mean mid-week, and not like the wild weekenders. Entering the ripe old age of Thirty Three, I surpass the life of Jesus. But, not in terms of his life's legacy. Still, I start my second adulthood. A new chapter in the creation of the Griffins! Born, once again, and my family walks right beside me as youth will not be wasted, this go around.

I promised Mom that I would spend this Birthday with her. I am normally not in New York on my Birthday. Ugh! New York is the pits and I can't wait to move the fuck on. But, for now Mom and I call one each other, wishing and praying that the other one will cancel.

When I arrive at my home station, I do wat I always do. I look upward and East as I make the short walk home. Up Jackson Avenue, I look up at the Dark Sky that is clear as the night is long. I notice that the Stars appear to be moving along with me. I think perhaps the stars only, appear to be moving along with me. I've never noticed or found evidence that the Stars move. Don't they always seem to stay in their places as we move about?

The stars closest to Earth, almost like the height of the phone lines and tallest buildings, seem like Jelly Fish tredding water. I see Two, or maybe Three little creatures floating, almost as if they were walking on the tight ropes of a circus act. They're still far enough from me that I can't fully make out the formations of what is mirroring my footsteps, yet they're above me.

If I had to describe them, I would say that the perfect description would remind you of those little Christmas Tree Angel ornaments. You've seen the ornaments, where the little angels have wings and hold a trumpet at their mouths. Yet, these guys all have legs, which movements appear to be spinning like the wheels of a vehicle. They are glowing neon colors Blue, Green, and Red. Not in a mixture, but solidily, and individually.

As I look further across and up, I see even more of these things. I ask, "What the hell are those? Ailiens?" Then God replies, "I told you that you would see him again." "You mean…?" I try to get out, before God interrupts, "Yup. They're Jeremiahs." No fucking way!

You gotta be kidding me! All this time, as I bounce around the city, ya'll mother fuckers are hovering above me? Seeing my every move with your physical eyes? And, all of the car horns and external sounds that I use to validate what I hear, you guys were manipulating physically as you fly above it all.

As that slips out, they begin making car horns go off, from the cars parked on the street, and not a single soul is in any car. They even make one bad ass, Toyota 4Runner start up all by it self. I look around to see if it was possible for the owner to be approaching, after starting the truck with a remote. But, nope. Us alone, communicating with one another. What a great Birthday treat. Jeremiah kept the secret long enough for me not to have ever noticed them before now.

Again, I wonder if I am the only person that can see these Angels above us. No one else on the street seems to be taking notice to them. Or perhaps since I am the one who has only arrived, they're already use to them.

I linger on my stroll just for a short while. As I marvel at God's creation, there in plain sight, right in front of me. And, for everyone else to see, I think? I'm not sure you would call it psycosis, but I feel as though something has changed in me. I sometimes wonder if when I made it to the hospital in Staten Island, and they put me under with an "I.V.," if they really injected something into my body, or worse. They may have operated on me once I past out. Whatever the case is, at night when I look at lights. The lights tend to starburst around the edges of the light fixtures. Sometimes all I see are Angeles that should be fixed upon the top of a Grand Christmas Tree.

How can I, now all of sudden, only see Angels when I look up, in to the night's lights? If I look up into a window, and I see some one flick on the light, I'll pause for a moment, thinking, there goes Jeremiah. And, then I come to my senses and say, no boy. That's only a light.

Jeremiah was so over me that he even deleted me as his Facebook friend. As I look at his page now, I am prevented from seeing videos of his Homegoing ceremony. It is probably for the best. Who knows if seeing something so representative of his end, could drum up more mourning and negative emotions toward a situation that, I for the most part, feel fine about.

Understanding that no one ever dies, but we transistion. Which, is something I suspected long before Jesus showed up. But, since he has come to validate that idea for me. I now know that some of them, like Jeremiah won't let go of their social media, from the other side.

Eventhough he won't add me as his friend. I can tell when Jeremiah, or God, for that matter, uses social media to communicate something to me. I will sometimes click on the Facebook ap on my phone, while randomly, my address book will pop up. I'll be quick to curse the phone out, since the Two ap buttons are no where near one another on my phone.

Yet here I am looking at my address book, that oddly has oppend on the letter S. Instead of the letter A. I'll then look closely at the first name and number to be seen on my screen. Then I will hear Jeremiah tell me that I need to call the person that I have viewed. I get that he feels the person could help me. But, Jeremiah is not specific about what to call the person up for.

We don't really have in our process any room for fundraising over the phone. So, we'll have to be clever in why we call up this stranger from Facebook. Now when I get new Facebook friend request, I typically wonder Two things before I accept or ignore the request. Is this someone being guided to me by Jeremiah? Or, is this person infact dead, and trying to communicate with the one chosen for communicating between worlds. I mean, if there is such a thing that exists called Dead?

We're getting better with the power struggle. Jeremiah expects for Mark to accept his every suggestion. If I ever turn down his gesture, he makes me shake. It's kind of scary at first. But once it subsides, it feels like rocing a child off to sleep.

Granted, making Jeremiah my slave. Forcing him to work on things that have no intrest or benefit to him doesn't make me feel like I am living in a New World. But, his way or the highway is why I am so lonely now. I'm sure that God was expecting that Jeremiah and AC would naturally make me feel the same. But, AC and I were expanding together. Now as Jeremiah expands beyond where we transisioned, I feels stuck. Stuck in the same, old, familiar place, which lacks evidence of Jeremiah's happiness. That limits my happiness, and I've always agreed that each participant in a relationship should be responsible for their part of the happiness.

This experience of witnessing debates between Omhmar and Jeremiah is odd and brilliant at the same time. I know that I've become someone else, because Omhmar would normally do whaterver Jeremiah wanted. But, I focus on our bottom lines and the Grand Design. One time, Omhmar and Jeremiah met up at their favorite starting place in West Hollywood called the Abbey. This was many years ago, of course.

Jeremiah played with the colors of his hair, but for pretty much all of his life he only had Two styles, short or long. But, Omhmar's hair styles ran the gammot of any style you could ever see on a man or woman.

This particular night, Omhmar spent an hour flat ironing his hair to a flawless "Grace Jones" kind of hightop fade. Then, he curled the edges over to look like a tidal wave. He was so happy and proud of his creation, upon being the first to arrive at the Abbey. Omhmar made a lap around the bar, saying hello to all of his fans, and feeling like "The Shit."

About Ten minutes later Jeremiah arrives and tells Omhmar that he wasn't feeling Omhmar's hairstyle. Omhmar saunters off, through the dancing crowd, skips across Santa Monica Blvd. to his car, then books it home. Leaving Jeremiah standing there.

Since, Mom cancelled on me at the last minute. I could not find anyone to replace her sitting beside me at tonights New York Philharmonic's, Handel

Messiah. On the ride downtown I had a feeling that, God had intended for Jeremiah to accompanying me to the performance. But, I didn't expect to actually see him. I forget that their flying above me. Always with me. Where direction condition intersect.

When the moon isn't visible, they exist much higher in the sky, appearing to be stars lighting the night sky. They're closest when we can see the full moon. That's when he looks like an angel with wings and a trumpet.

Jeremiah once told me of a time when Beyonce attended a performance of Alvin Ailey. The dancers all noticed that her hair would flutter, as if Beyonce was being fanned while sitting their watching the performance. Tonight as the quiet boast of the Handel score is being performed to absolute perfection, I feel that wind. It circled around me often as I took it all in. You could really feel that the Christ Family was all present. Complete. To celebrate the Glory of God, and to.

Happy New Year! Mom and I plan each year to watch the ball drop in Times Suare, but never stay up long enough. During the winter, Mom keeps her house very warm and toasty. On the top of that, she only wants to watch *Praise the Lord* and other TBN preaching shows. There is no way I can keep my eyes open for all the hours until midnight. But, now we wake up in 2016 after completing the weirdest year of my life. It's called Saturn's Return. If you are in your late 20's or early 30's your Saturn's return is coming and I promise you that it will be just as wild. So, get ready boys.

After working out the kinks, and traveling through worlds. Right on January First, the Christ family and I began implementing the lessons learned during my training period, and psychic initiation. Although, we plan on launching East Coast Ballet Company until 2018, Jesus and I thought it would be good to produce a concert to validate my producing skills, and the choices that I agree upon, when designing Griffin Cleft's Five-year roll out. Together he and I make wish lists of sponsors, tour cities, home base cities, composers, designers, and many other important things needed to produce theater, opera, ballet, and film & T.V.

Jeremiah being an only child, and facts regarding the Stabat Mater, who bore Jesus, her only son, decided our theme. We agreed that it was time to mount a Griffin Cleft production. A Mother's Day Choral Extravaganza will be a collaboration of Guest artists, orchestral, and choral celebration of the Stabat Mater text. Performing Three different versions of the text, with East Coast Ballet Company dancing in One of the versions.

I could listen to any version of the Stabat Mater for weeks at a time. And, pretty much did for all of January through most of February. I'd change my mind often, like indecisiveness Libra do. I was sure of using the Pergolessi/Bach Stabat Mater. But, once the work evolved from the Baroque period, the works grew with larger orchestra and chorus parts. With Karl Jenkin's version, being the most recent and the largest of all. And, as equally fabulous, as well!

Around this time I began to get replies from the sponsors I'd submitted rquests to. The choices I made to support our efforts weren't as eager as I expected. Some how, I'm in the mindset that Jesus is trying to prove that the bond was his word. I only sent out proposals to the sponsors that we discussed. It is my normal practice to conduct aggressive fundraising initiatives appealing to hundreds of potential sponsors. But, I would never produce an event this large in only Five months. With no one on board to fund this performance that has now grown into my dearest dreams. I wonder what in the hell does Jesus and Them have up their sleves.

As March brings a close, to St. Patrick, it's time to begin filming the second season of *Mad Science*. When we began filming the first season, I had no idea that the "Mad" part of the title would mean that I was Mad at the death of Jeremiah. Nor did I know that the science would be him communicating from the "Other Side," leading me to clues that validates God's plan for us all here on Earth. My original idea was to try connecting some of the scientific facts regarding Human Evolution. Many passages in the Bible that I feel are exact depictions of the evolution to come.

Jamie and Adam do each week on *Myth Busters,* I wanted to dispute scientific claims to God or a higher power not being the creator of life, and evolution being it's Grand Design.

As cameras follow Jeremiah and I producing this Stabat Mater concert for

Jesus, Mary, and MaWest, I become very doubtful in our efforts. I thought that my producing chops would bounce back into my muscle memory. But, I was dropping the ball left and right. Topping the list of worry was funding the production budget.

I'm confident that my old donors in Los Angeles will be proud to hear of my return, and would probably right a check. But, that wasn't enough to drench my doubt. I began editing the budget, scaling it down to its simplest form that I could rest easy in producing. It's not an easy task since so many moving parts, and creative staff is required whenever performing live music. Especially with an orchestra and chorus. There is high costs for the venue that I let Jesus talk me into reserving. The world famous Apollo Theater in Harlem. I know!

Whenever I ask Jeremiah to way in on a decision, he always replies with some bullshit like you gotta believe. Or, it's gotta be your choice. Fuck! How can he reside in the place of infinite wisdom, and still wants to just show up on the gig date, leaving me to do all of the production work? That was our process when he was alive!

I'm trying to believe. But, Jesus' suggestions only pan out for me every other time. On the flip side, Jesus' suggestions pan out for me every other time. I can move forward with the odds being 50/50 in my favor. The guest artist I had scheduled to appear at this point weren't big enough to attract the sizable investor we needed. So, Jeremiah and I turned to the internet, searching for some A list opera singers I could add to the bill.

For MaWest and other Black Moms, I wanted to feature a couple of Black Sopranos. I knew it would be just the special touch they'd appreciate, along with inspiring their young family members to train in classical music. But, our search is a bust. Opera has the same inequality as Ballet does, in that they both lack appeal to minorities.

"Who's your favorite Opera singer?" Jesus asks me. "Cecilia Bartoli." I reply without hesitation. "But, who was it for many years of your childhood before you discovered any other?" Jesus re-asks. "Kathleen Battle!" I say with a big smile. But, she can't be available at such a late date. A diva like that is booked a decade in advanced.

"What if there was a cancellation in her schedule?" Jeremiah guides me, while my mind shifts through the cancellation process for someone of her magnitude. In almost every case, a cancellation can leave a diva without work for up to an entire year. Alternate job dates always conflict with the job that has already been booked for after the cancelled contract. After Jesus played me with those damn sponsors, I almost want to smother my hope. But, with a name like Ms. Kathleen Battle on the bill, I could raise the entire budget with a single budget. Not to mention sale out in minutes.

"Hello, this is Mark Griffin, managing producer at Griffin Cleft. I would like to check the availability of Ms. Battle a New York City venue this coming Mother's Day. Yes, I'll hold." Like the mercury of a thermometer, my blood is rising and lowering. My teeth chatter from my efforts not to smile. Jesus has a way of getting me to let my guard down, and then strengthening my faith without even realizing. And, I don't want to have too much faith in him, nor anyone else besides myself.

Still, some how. Some how faith and good health is all he seems to be able to compensate me with, for my ground breaking, cutting edge portrayal of the Chosen Ten. On top of my financial giving, and charity work, and all or my pro bono business consulting, life coaching, and community advocacy work. Which is, outside all the ridiculous spiritual and ritualistic things he makes me do. And, without question. "Yes, hello?"

Ms. Battle's Agent has returned with her schedule. "You're kidding me!" After discussing performance repertoire, fees, deal points, etc. I'd just submitted an offer to contract my childhood idol. So often I stop to speak aloud, "You've thought of everything." You know who I'm talking bout.

Normally, this is the part where I pray and worry. Begging the Universe to allow God to answer only my need, while making deals of sacrafice, and offering up praise. But, since he came to me this time, our tables are kind of turned. Jesus and Mary really want this concert to happen, and they've promised to pull out all of the magic required to perform miracles for the person chosen to document the evidence of such miraculous times.

Promises, promises. Loud as day, all day, and every day. Ringing throughout my dreams, proclaiming that Jesus is real. My Lord and Savior. Who hasn't forgotten what he owes me. But, while he waiting for some checks to clear, I have to spare him a dime, some time, and the use of my mind.

Thank God! Some comic relief to prevent me from following my trash down the incinerator has just touched down at JFK Airport. It's been Seven years since I have seen the "The Future Superstar of the World!" Yes people, he not only called himself that when making an introduction. But, Mr. Ray Cruz actually had it printed on his 8x10, Black and White Actor's Headshot! I kid you not. My childhood friend, who shared a ride with me out to Califronia, during our first migration West. He stayed in Los Angeles, as I moved back to New York. Irronically he hates Los Angels as much as I hat New York. We've been talking about swapping cities, to live in our preferred cities.

Any of my friends that successfully meet the criterium of a Bueen, receive imperial customs as our Bueenship welcomes them into our Bueendom. I roll out the aristocratic courtesies, along with deeming us all with royal titles. As Jeremiah and I wait for Ray to arrive in front of my entrance. I wonder which family title and estate I should bestow upon him. We haven't seen each other in nearly a decade, and he may be a whole new person. I perceive how difficult it is for people who knew Omhmar, to accept that Mark is someone entirely different. Not wanting to judge a book by its antiquated façade, but if memory serves me correctly. I think that I shall name him Baroness Rayford Fela Cruz, of the Griffin/Cruz Manor Estate in Source City, Marks. Hope it suits him.

This looks like him approaching. "Sound the Trumpets" Jeremiah instructs the Court Musicians that have prepared a special welcome march by our Court Composer Henry Purcell, in celebration of Baroness Rayford's safe arrival. "Oh my God!" Fela Cruz belts from a half a block away. His pipes are as powerful as the ones ringing by our Court Musicians. And, it looks like he didn't come alone.

"Oh my God1, I can't believe that you look exactly the same. Bitch, you could have at least gained some damn wait, or something." "A year and a half I have managed to reside in Bueensboro nearly un-noticed. You've been here Ten seconds, and have commanded that the entire compound take notice to you." I reply to his albatross of a greeting.

"What an adorably sweet, little pooch you have here." I compliment his traveling companion. "This is my baby, Celia. Celia Cruz." He presents the most Blue Ribbon worthy of a Yorkie Terrier to me. I would present my royal court and employ to him, but he can't see them, and would probably call me crazy. Maybe even fear staying with me, in my castle. Instead of trying to explain how I have dual citizenship in both reality, and in the future that you must be chosen to see. I ignore the curious stares of the towns people, grab his luggage, and into the Griffin Cleft Tower. On the ride up, I wonder if Celia is more suited for the title of Baroness than Fela Cruz is.

"Your place is too cute!" Lord Rayford tells me. I'm feeling more like Ray should be a Lord. Or, maybe Mother Ray? "Are you kidding me? I reply to his concerns regarding how much money I spent on the place. Isn't it funny that you can feel what a Grey accomplishment someone has achieved, and then when they tell you that they have spent guap on professional help? All of sudden, the fantastic accomplishment dimmenishes, and finally we end up saying, "Girl you got jipped!"

As Mother Ray saunters about the headquartes for Christ Inc., aka Griffin Cleft-NYC. Better yet, Mother Rayford. I breathe him all in. I evaluate his energy. Unfortunately he's one of those guys that you thought still lived with his parents, or on somebody's couch, until you find that he's actually wealthier than anyone you know. When he speaks, you hear not only his undisputal faith and true belief, but also his Ivy League education. His father, a successful Pastor of one of our Nation's Capitol's, most prestigious home of worship, paid a mint so that Mother Rayfor could sound both like Madea, and Jackie O.

I've tallied my evaluation, and can assert that he is doing just fine. Better than I infact, although his compliment that I look exactly the same resonates in multi-dimesions. I feel the same, and stuck! I left Los Angeles as a Dancing Business Man. I know, a fucking circus, right? And, here I am now a Dancing Fool, trying to convince myself hat Jesus, my deceased Nephew and Best Friend are working from separate worlds to ensure that I, as Mark Griffin continue the work that implements the Grand Design that God chose Omhmar Griffin to do.

Then my sister's father, whom raised me as well. But, we did not develop any relationship past the point where he and Mom split. I was only ten years old then. Now, Twenty years later, King was able to convince everyone that he would be the best History and Theology professor for Mark, providing the best education possible to aid his transistion to Mark's World. Which, Jeremiah and I have decided to to name the world, Marks. AC already told me that the Planet Mars was to become my world. And, that I named it long ago. But, the Spiritual relm that decompressed the newest world at that time, did so during a period where everyone thought it to be bad luck to use the letter K.

Therefore, versions of Mark since then, have now become Marcus. Not to mention that, now that Mark has entered his world, that we've all been building for him. "They forgot the K!" Is what he says, in regards to the titling of his planet Mars.

Jesus and I tried keeping our circles small. My perception of the Christ family is that, we're the Democrats of the Spiritual world. I'd imagine that Heaven is probably as lovely as they claim it to be. I've seen small performances of my God, and I am sure that he has spared no expense for us. However, I have also witnessed the effects of having to dumb down life for the many of Christians who are just plain dumb.

So, it would not surprise me if, similarly to this year's Rio Olympic Games. Gorgeous pageantry, and corteous welcomes are in store for all the beings lined at the Pearly Gate. And, you'll all get to relax, as soon as my father in Heaven, figures out where he put that key. Bear with him. He'll look insistently for the key before remembering that h'd never need a key, within his abundance of powers.

He claims that he has given all the same power to inact on Earth. So, The Purple Period, and Purple Book. The Journey, and experience. And, God's Process, is all for anyone that shall believe, to gain the powers readily accessible to us in our Spiritual tool box. And, for me to become self aware.

A lottery, un parallleld to another. Yet, many of us, perhaps even you as you sit there reading this. Submit to the concept that life shouldn't be filled with more than the stuff that makes us who we are, only on the surface. A poet. A mother. Or, phenomenal in the sack!

Keeping in mind that I have been detached from all of the life that use to be only, Omhmar's. All of his friends, family, and favorite locales have been denied acces to me, after making the transition to becoming Mark. Now that Count Rayford has arrived with all the latest "Words on the Street." I feel as though I have no clue who these people are in the stories that he presents me with. Stories that should be calling my memory forward. Instead, I enjoy the colorful depictions of Omhmar's friends, as though I've never met them.

Funny enough though, he tells me that he has recently run into a friend that I know was murdered. I almost don't believe him. Yet, I must when I see the recent picture added to his Facebook profile, of the Rayford, and the deceased friend standing side by side. Could this be even more validation that I truly have been moved to a world that I did not exist in. The other side, where no one dies. But, we all simply transition. If we come from Source energy and energy is never destroyed nor lost, but trnasfered. It is very well possible. Also, remember that God instructs me that I can believe whatever I want to believe, of what I think I have heard. Especially, when I think that the source of the information that I've heard, has come from my Father's mouth, himself.

Some of the new truths that I discover in the Purple period are so in line with what I originally wanted to disprove in our Status Quo. And, had no resistance to included them in my belief system. But, I always thought that our transistion would be to another place, totally removed from anything that is like Earth as we know it. The idea that we can also transistion to a world identical to the Earth that we just left blows my mind.

It is probably the single bit of information that needs a bit more chewing before I can digest it entirely.

So often, through out this experience, I want to share what is happening in my life with the people closest to me. But, I frequently stop myself, eventhough God tells me that there is not a piece of it that I can not share. However, getting the people whom never been chosing to know the truth, to actually believe and take the contrasting evidence into consideration that disputes so much of our truths is quite an uphill battle.

In fact, many of you who our now reading my collection of evidence, will deem me as crazy! Don't worry yourselves. Since, God has already told and prepared me for a large number of Humans to label me with such an insult. Though, at times I would agree with you. However, there are small numbers of you who have beeen chosen. The decision has not been made lightly. And, the Laws of Attraction has now placed you in this very moment, in front our Purple Book, challenging you to now stop your hesitation. And, just believe.

Others won't need as much convincing. When I exammining some moments from the times that Lord Cruz and I lived in LA. I have often arrived at the same conclusion. That Jesus must have chosen him before. As he shares his industry's best gossip, I share with him some of my difficulties stemming from being chosen. I guess I've recalled, to his memory, now some of the times we had. The stories from our past never change. Yet, this time their meanings are inclusive of many more valuable components.

He and I moved to Los Angels on a cross country trip from New York, when we were invited by his best friend who would be driving. The Three of us piled into a Two door Camry and set off from Staten Island to California in 1999. Which inspired Omhmar to get the Four door Camry as his first "real" car. We're speaking about cars, because I have brought his attention to how the cars here on Saturn are very new and different then what and impoverished community normally parks at its door steps.

Everyone including God, Jesus, and The Illuminati all requested that I make decisions regarding the rewards that I would receive for my hard efforts.

Consistantly I have asked for a car for Mom. I don't need wheels in New York City. In fact owning a car in this city is a very inconvenient and expensive property. If you actually arrive at your car that has not been stolen or towed for parking violations. You are stuck in traffic, wishing that you had just jumped on the Brooklyn Bound Four Train. What he can not know, is that most of the cars have either a relation to Jeremiah, or are cars included in my wish list of rewards.

Even the colors, year, make, and the numbers on the liscense plate, all correlate to discoveries and findings that I have included in my Log of Evidence.

Similarly to the overhaul of Mantanence and Management staff of my apartment complex. On a visit to Mom's house, I've noticed that all the homes on her block have been sold to owners who have just moved in. Every single home, besides Mom's. And of course, her best Golden Girl, Nina. Which is not actually her name. But, Mom insits on calling people by the name she wishes to call you. Nina lived with her daughter's family, directly across the street from Mom. Added to that, Mom's home attendant also changes on the first day that I visit her on Saturn, while noticing the shift in neighbors.

Mom's new neighbor that became her ride to church, drives Mom to church in the exact same Mercedes that I ordered for Mom, during the time I spent with Skull and Bones.

After I was invited to participate in the validating election, an arrangement was agreed upon. Which allows me to aid the Trump Campagin, with a large cash salary, plus that same White Mercedes that I plan to give to Mom. However, after strategizing Trumps best options, and pitching the strategies to Trump's Campagin Headquartes. I have gotten no response. Why am I supposed to believe that God can do anything? And, he can not even get the Trump Campaign to respond to my proposal.

Additionally, without seeing any validation from Trump, himself. I still have to believe that Trump knows who I am. That he is a true, chosing believer. And, that my Spirt Guides have made all the arrangements to unite Trump and I's Greater Good causes, for a successful win, this

Presidential Election. Moments like this is really when my faith shifts, to lean more towards the doubts of me being chosen for anything. And, that I am really becoming Skitzo, like my Uncle Gordon.

Other moments, now that I've returned to the world where Trump and Clinton are paired in a fierce battle, really give me doubt.

I became a Republican so that I can vote with my Party in the Validating world. A shocking decision for the entire Spirit World that made the assumption that I would chose Clinton. Perhaps that's because they knew what lay a head in Trump's laughing stock of a campaign. And, now that my pick has shed light on what he really stands for, has really promoted disbelief in God's Grand design.

It really seemed like God and I were in agreement on Trump being the best candidate. Why would he even put Americans in the position that would even have to consider this ass? Nevermind, actually agree that he is the better choice. Which, is based off of criteria that I can not share here. My own purpose for not wanting Clinton to win is simple, and maybe even superficial, however. I felt that America celebrates its independence in July because America was able to defect from the status quo, of being controlled by a single Monarchy.

When Clinton wins, the familes to have power in the White House will have included only the Clintons, and the Bushes for Sixteen Years. I know that I could have just spoiled the surprise. But, I purposely waited to publish my Purple Period findings, until after the election. So, that I would not influence or manipulate anyone into letting their vote, be my voice. But, yes. I heard that Clinton won, and America will finally have its first Female Peresident, a year from now. Undortunately, I have to keep the secret. Which makes me look at every election event that's covered, as if it's bullshit.

I understand that God never said that our choice would be the winning choice. Instead he has asked that I begin a process that allows for me to keep the commitments belonging to my decisions. As a result, the first time that I decide on a candidate, before everyone else has, I happen to choose the looser. Which is thrilling to me since, I wish that I didn't have to vote for him. But not only am I stickler for voting with my Party, but also because. "You are not voting now, you have already voted. And, your vote was for Donald Trump." God, reiderates.

I'd already made the decision that the best way that I could help to make this election a fair election, was to stay out of it. I wanted to stand on my soap box, yelling from the top of my lungs. For all to hear the reasons why I feel as though now is the time for more minorities and Democrats to make the swithch to the Republican Party. However, after seeing that dark cloud hovering over America's that we don't want either candidate to replace President Obama.

As far as being the first True Christian Man to marry another Man? I feel as though that ship has passed. Many Christians have gotten married to the same sex, since the Supreme Court ruling. Jesus asks "How do you know that any of the married couples are True Believers?" I have no clue! No clue of what metric differentiates that True Christian from the Pretending Christian. Yet I will continue my search, until I hear, "it is too late." Again.

I wish I had better success to document. But, as the Purple period quickly fades to its end. I am full of confusion, doubt, disappointment, and feel as though I have been left alone on Earth in a far worse position than I was in before Jesus or anyone else showed up. Did I really Meet Jesus. Was I really Chosen? Am I just loosing my sanity? Or, have I just fallen for the simple manipulation of the Desguise Master?

In an effort to make since of all that has happened, I have not only decided to seek clinical therapy. But, I seek such therapy for the purpose of inacting an aupothasy from the Church, and the Doctrine of Jesus Christ. Yes he's real. Yes he does see all, and firmly stands behind his teachings. I have much evidence and validation for what we learn as kids, from the Bible. His Father is both miraculous and mysterious. However, I just can't seem to find a purpose for any of it to matter to me. Nor, any of you.

As I sit here trying to produce this major Choral Extravaganza Mother Mary, and the Mothers of our world, this Mother's Day. I realize that I really am not Omhmar Griffin. I am learning minor production skills, that should be able to be accomplished in my sleep. Yet, It takes me Three times longer than the time I suspect it would take Omhmar to accomplish the same tasks. When Jesus presents the rare possibility of my favorite opera singer, Kathleen Battle being available I run with it. He was absolutely right, Ms. Battle is available during our production dates. However, she lacks the interest to work with me. As I have no reputation as Mark Griffin. After offering her more and more money, and she still declines my offer, I have lost Three weeks in my pre production, waiting for her answer.

Omhmar would have never waited Three weeks to get an aswer form a single artist. He never even made an offer one at a time. He would choose several artists that can be cast in the roll, that he seeks to cast. Then, he would make offers to all the artists on the same day. Then, the roll would go to the first artist to say yes. However long it took to cast, he at least had a firm yes from somebody.

But, here I stand with Three months left to produce my first major event as Mark Griffin with Truth Hurts, as the only headliner. An R&B singer to star on a Classical Music bill. Not to mention, that I booked the world famous Apollo Theater in Harlem for Omhmar's return to the limelight.

New Yorkers are a very forgiving audience. Since Jesus chose me, he has brought me nothing but lies, confusion, and chaos. Not to mention taking some of my favorite loved ones from me in the process.

Clearly there is still a great deal of practice and preparation that I need to conclude before moving forward in God's process, or. For me to step out of Omhmar's shadow and become my own celebrated American Producer. And, I still have to find a husband, when I never even wanted to get married, in the first place! I stood firmly in my position, that marriage rights should be reserved for Heterosexuals. Homeosexuals should not have to be forced to now participate in a religious practice that was designed, specifically with our exclusion as part of it.

I will continue to explore my findings. Enroll in an intense mental rehabilitation program with a leader that enjoys a challenge. An Aupothasy from Christianity is not only much needed. But, Mom can no longer be in my ear beating the drum of my need to get my life right with the Lord. I think the Lord and I are good. Most importantly for me to do right now, is head to Ma West, and let her know what her sons are up to. I'll tell you what happens next, when I talk to you next.

The End

ABOUT THE AUTHOR
OMHMAR + JEREMIAH = MARK MOZART

Omhmar and Jeremiah were best friends, to say the least of it. Professional dancers on stage and screen, the pair were almost identical twins. Omhmar was raised by a Single Black Mother in New York City. Jeremiah was raised by a single Black Mother in Long Beach, California. They've both shared spaces all over the world. Now they share one body. That body is called Mark Mozart. And, you have just read the story of their first Transistion.

www.ingramcontent.com/pod-product-compliance
Lightning Source LLC
Chambersburg PA
CBHW071416180526
45170CB00001B/126